THE CONSTRUCTION OF BUILDINGS

VOLUME 3

FOURTH EDIT

By the same author

Volume 1
Foundations and Oversite Concrete –
Walls – Floors – Roofs

Volume 2
Windows – Doors – Fires –
Stairs – Finishes

Volume 4
Foundations – Steel Frames – Concrete Frames –
Floors – Wall Claddings

Volume 5
Supply and Discharge Services

THE CONSTRUCTION OF BUILDINGS

VOLUME 3

SINGLE-STOREY FRAMES, SHELLS AND LIGHTWEIGHT COVERINGS

R. BARRY
Architect

FOURTH EDITION

OXFORD
BLACKWELL SCIENTIFIC PUBLICATIONS
LONDON EDINBURGH BOSTON
MELBOURNE PARIS BERLIN VIENNA

Blackwell Scientific Publications
Editorial Offices:
Osney Mead, Oxford OX2 0EL
25 John Street, London WC1N 2BL
23 Ainslie Place, Edinburgh EH3 6AJ
238 Main Street, Cambridge, Massachusetts 02142, USA
54 University Street, Carlton Victoria 3053, Australia

Other Editorial Offices:
Librairie Arnette SA
2, rue Casimir-Delavigne
75006 Paris
France

Blackwell Wissenschafts-Verlag GmbH
Düsseldorfer Str. 38
D-10707 Berlin
Germany

Blackwell MZV
Feldgasse 13
A-1238 Wien
Austria

First published by Crosby Lockwood Staples 1963
Reprinted 1967
Second edition published by Granada Publishing in
Crosby Lockwood Staples 1972
Reprinted 1974, 1976, 1979
Reprinted by Granada Publishing 1981
Third edition 1984
Reprinted by Collins Professional and Technical Books 1987
Reprinted by BSP Professional Books 1988, 1991
Fourth edition published by Blackwell Scientific Publications 1993

Set by DP Photosetting, Aylesbury, Bucks
Printed and bound in Great Britain by
the University Press, Cambridge

DISTRIBUTORS

Marston Book Services Ltd
PO Box 87
Oxford OX2 0DT
(*Orders:* Tel: 0865 791155
Fax: 0865 791927
Telex: 837515)

USA
Blackwell Scientific Publications, Inc.
238 Main Street
Cambridge, MA 02142
(*Orders:* Tel: 800 759-6102
617 876-7000)

Canada
Oxford University Press
70 Wynford Drive
Don Mills
Ontario M3C 1J9
(*Orders:* Tel: 416 441-2941)

Australia
Blackwell Scientific Publications Pty Ltd
54 University Street
Carlton, Victoria 3053
(*Orders:* Tel: 03 347-5552)

British Library
Cataloguing in Publication Data

A catalogue record for this book is available from the British Library

ISBN 0-632-03742-3

Library of Congress
Cataloging in Publication Data

Barry, R. (Robin)
The construction of buildings.
Vol. 3 published: Oxford; Boston:
Blackwell Scientific Publications.
Includes indexes.
Contents: v. 1. Construction and materials (5th ed.)—v. 2. Windows, doors, fires, stairs, finishes (4th ed.)—v. 3. Single story frames, shells, and lightweight coverings (4th ed.)
1. Building. I. Title.
TH146.B3 1980 690 81-463308
ISBN 0-632-03742-3

CONTENTS

PREFACE

For this new edition a major rearrangement has taken place to accommodate the most recent developments in the construction of single-storey buildings. The functional requirements of the various construction elements are given more prominence. The use of new materials, such as deep profiled, trapezoidal, plastic coated sheeting and single-ply roof membranes, are covered and new construction methods, such as tilt-up construction, are introduced. The important areas of the influence of enhanced insulation on roof shape and the provisions for moisture vapour checks are covered.

Chapter One has been revised to describe the development of the single-storey shed frame. It covers the early closely spaced triangular truss frames, suited initially to slate roofing, and that were for many years adapted to suit corrugated iron, steel and asbestos cement sheeting, through to the more widely spaced lattice beam and portal frames suited to the low pitch, trapezoidal section steel sheeting that is commonly used today. The development of cold formed purlins, lattice purlins and solid web I-section purlins for widely spaced structural frames is described, together with wind and structural bracing necessary for stability. The comparatively recent use of composite frames of precast reinforced concrete columns with lattice steel roof beams for low pitch roof coverings is also considered.

Roof and wall cladding, thermal insulation and flat roof coverings, previously included in separate chapters, have now been consolidated in Chapter Two to embrace the functional requirements common to roof and wall cladding. The development of profiled cladding from the early shallow section corrugated iron, steel and asbestos cement sheeting to the now generally used trapezoidal section, plastic coated metal sheeting, including standing seam sheeting, together with thermal insulation, form the main part of Chapter Two. The development of flat roof weathering from asphalt and fibre felt to high performance and single ply weathering is described and illustrated.

Chapter Three considers the functional requirements of rooflights and describes the use of glass and plastic sheeting.

Chapter Four considers the original material covering diaphragm and fin wall construction together with a description of a recent innovation in the use of tilt-up reinforced concrete wall panels as wall enclosure and support for lattice beam roofs as a form of rapid, solid construction for single-storey buildings.

Finally, Chapter Five deals with shell structures.

INTRODUCTION

The small-scale buildings described in Volumes 1 and 2 are constructed with traditional materials such as brick, timber, slate, tile and non-ferrous metals that have been used for centuries and have stood the test of time. The useful life of such buildings, if reasonably well maintained, is up to 100 years or more.

This volume, Volume 3, describes the construction of single-storey buildings, such as sheds, warehouses, factories and other buildings generally built on one floor, which account for about 40% of the expenditure on building in this country.

Over the past 50 years most single-storey buildings have been constructed with a structural frame of steel or reinforced concrete supporting lightweight roof and wall coverings to exclude wind and rain and to provide insulation against loss of heat. The small imposed loads on roofs can be supported by thin, lightweight sheets fixed to comparatively slim structural frames to provide wide clear spans between internal supports. The thin, lightweight materials that are used for economy in weight and first cost are not robust and do not withstand for long the destructive effects of weather, dimensional changes and damage in use that occur in buildings.

The consequence of the adoption of lightweight materials for roof and wall coverings, for the sake of economy, is that many single-storey buildings have a useful life of only 20 to 30 years before considerable works of repair or renewal are necessary to maintain minimum standards of comfort and appearance.

The concept of functional requirements for the elements of building is now generally accepted as a necessary guide to the performance criteria of materials and combinations of materials used in the construction of the elements of building. In traditional building forms one material could serve several functional requirements, for example, a brick wall which provides strength, stability, exclusion of wind and rain, resistance to fire and, to some extent, thermal insulation.

The materials used in the construction of lightweight structures are, in the main, selected to perform specific functions. Steel sheeting is used as a weather envelope and to support imposed loads, layers of insulation for thermal resistance, thin plastic sheets for daylight, and a slender frame to support the envelope and imposed loads. The inclusion of one material for a specific function may affect the performance of another included for a different function which may in turn necessitate the inclusion of yet another material to protect the first from damage caused by the use of the second material, for example, where a vapour barrier is used to reduce condensation on cold steel roof sheeting.

Recently the demand for space heating and the consequent inclusion of materials with high thermal resistance has led to problems in building unknown to past generations who accepted much lower standards of heating and more ventilation of their homes and work places. The inclusion of layers of thermal insulation in the fabric of buildings, to meet current regulations and expectations of thermal comfort, has led to the destructive effects of condensation from warm moist air and also to the large temperature fluctuations of materials on the outside of buildings which has been one of the prime causes of the failure of flat roof coverings.

The use of thin, lightweight materials for the envelope of buildings has been for the sake of economy in first cost with little regard to the life of the building or subsequent maintenance and renewal costs. Where the cost of one material used in the construction is compared to the cost of another, account should be taken of the relative costs of the elements of typical buildings as a measure of the value of saving. A guide to the comparative costs of the elements of single-storey buildings is as follows:

- Drainage and works below ground 15%
- Structural frame 15%
- Cladding including windows and rain-water goods 35%
- Floors and finishes 10%
- Heating, electrical and other works 25%

NOTE ON METRIC UNITS

For linear measure all measurements are shown in either metres or millimetres. A decimal point is used to distinguish metres and millimetres, the figures to the left of the decimal point being metres and those to the right millimetres. To save needless repetition, the abbreviations 'm' and 'mm' are not used, with one exception. The exception to this system is where there are at present only metric equivalents in decimal fractions of a millimetre. Here the decimal point is used to distinguish millimetres from fractions of a millimetre, the figures to the left of the decimal point being millimetres and those to the right being fractions of a millimetre. In such cases the abbreviations 'mm' will follow the figures, e.g. 302.2 mm.

R. BARRY

CHAPTER ONE

LATTICE TRUSS, BEAM, PORTAL FRAME AND FLAT ROOF STRUCTURES

Up to the latter part of the nineteenth century the majority of single-storey buildings were of traditional construction with timber, brick or stone walls supporting timber-framed roofs covered with slate or tile. The limited spans, practicable with timber roofs, constrained the rapid expansion of manufacturing activity that was occurring during the nineteenth century to meet the demands of the rapidly increasing population of England and the very considerable export of finished goods.

The introduction of continuous hot-rolled steel sections in 1873 led to the single-storey shed frame form of construction for most new factories and warehouses. This shed frame form of construction consisted of brick side walls or steel columns supporting triangular frames (trusses) fabricated from small section steel members, pitched at 20°, to support purlins, rafters and slate roofing. The minimum pitch (slope) for slates dictated the shape and construction of the steel roof trusses. This simple construction was economical in first cost in the use of materials, light in weight, easy to handle and quickly erected to provide the limited requirements of shelter expected of such small structures at the time. A symmetrical pitch single-bay shed frame is illustrated in Fig. 1.

The introduction of corrugated iron sheets in 1880 and corrugated asbestos cement sheets in 1910 made it practical to construct roofs with a minimum pitch of 10° to exclude rain. Nonetheless the single-storey shed frame of triangular trusses pitched at 20° continued for many years as the principal form of construction because of the simplicity of fabrication, economy in the use of materials and speed of erection.

Natural lighting to the interior of shed frames was provided by windows in side walls and roof glazing in the form of timber or metal glazing bars fixed in the roof slopes to support glass. To avoid sun glare and overheating in summer the north light roof profile was introduced, a light section steel roof truss asymmetrical in profile with the steeply sloping roof fully glazed and facing north. A single bay, north light shed frame is illustrated in Fig. 1.

With increase in the span of a triangular roof truss the volume of unused roof space and the roof framing increases and it is, therefore, of advantage to combine several bays of the shed frame construction to provide cover with the least volume of roof space and roof framing. To minimise the number of internal columns that would otherwise obstruct the floor, the 'umbrella' or cantilever roof was adopted. Lattice girders constructed at mid span in each bay support the trusses and widely spaced internal columns in turn support the lattice girders to provide maximum unobstructed floor space (Fig. 1).

The flat roof form of construction for single-storey buildings was, to a large extent, for appearance. The beam or lattice girder grid on columns affords no advantage in unobstructed floor space and little reduction in unused roof volume over the umbrella roof. The clean flat roof line and strong horizontal emphasis was accepted at the expense of many failures of flat roof coverings. In recent years improvements in materials and detailing of junctions have gone some way to repair the ill-repute of flat roofs. A typical single-storey flat roof frame is illustrated in Fig. 1.

With the improvements in cold metal forming techniques that were developed towards the middle of the twentieth century, a range of deep profile steel sheets was produced that could be used as a successful roof covering at a pitch of as little as 6°, so that the traditional triangular roof truss frame was no longer the most economic or satisfactory form of roof frame. It was then that the low pitch portal frame and the low pitch lattice beam frame came into general use to combine the benefits of low pitch profiled sheetings and the consequent reduction in unused roof space to be heated and insulated.

The plastic theory of design proposed by Professor Baker led to the use of the rigid steel portal frame for single-storey buildings. The rafters of the portal frame are rigidly connected to the posts in the form of a slender frame that is free of lattice members and can most economically have a shallow pitch suited to the profiled steel roof sheeting and decking that came into production in about 1960. Figure 1 illustrates a portal frame.

To reduce the volume of unusable roof space that has to be heated and the visible area of roof, for appearance

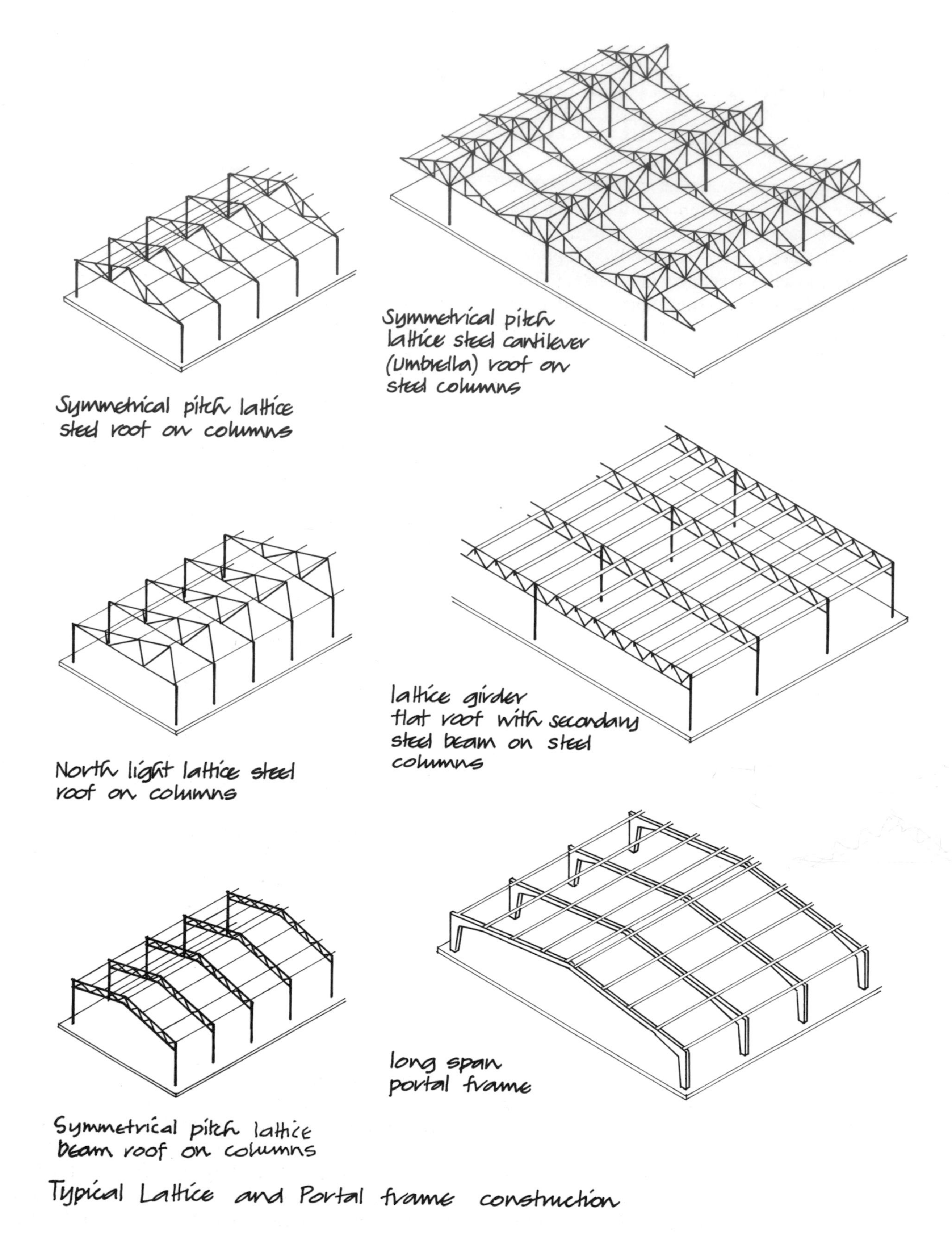
Symmetrical pitch lattice steel cantilever (umbrella) roof on steel columns
Symmetrical pitch lattice steel roof on columns
lattice girder flat roof with secondary steel beam on steel columns
North light lattice steel roof on columns
long span portal frame
Symmetrical pitch lattice beam roof on columns
Typical Lattice and Portal frame construction

Fig. 1

sake, it has become common practice to construct single-storey buildings with low pitch roof frames either as portal frames or as lattice beam or rafter frames. A lattice roof frame is fabricated as a lattice of small section members welded together in the form of a beam cranked (bent), a symmetrical pitch roof or as a multi-bay butterfly roof with valley beams, or as a single or double very low pitch beam. A single bay symmetrical pitch lattice beam or rafter roof is illustrated in Fig. 1. With the recent introduction of long lengths of profiled roof sheeting with standing seams that can provide cover in one length without end laps at a pitch (slope) of as little as $2\frac{1}{2}°$, the lattice beam roof frame has become the most commonly used form of roof structure for most single-storey buildings.

FUNCTIONAL REQUIREMENTS

The functional requirements of framed structures are:

Strength and stability
Durability and freedom from maintenance
Fire safety.

Strength and stability

The strength of a structural frame depends on the strength of the material used in the fabrication of the members of the frame and the stability of the frame or frames on the way in which the members of the frame are connected, and on bracing across and between frames.

Steel is the material that is most used in framed structures because of its good compressive and tensile strength and favourable strength to weight ratio. The continuous process of hot rolling steel and cold forming steel strip produces a wide range of sections suited to the fabrication of economical structural frames.

Concrete has good compressive and poor tensile strength. It is used as reinforced concrete in structural frames for the benefit of the combination of the tensile strength of steel and the compressive strength of concrete, and the protection against corrosion and damage by fire that the concrete gives to the steel reinforcement cast in it.

Timber is often used in the fabrication of roof frames because it has adequate tensile and compressive strength to support the comparatively light loads normal to roofs. Timber, which can be economically cut, shaped and joined to form lightweight roof frames is used instead of similar steel frames for the sake of economy, durability and ease of handling and fixing.

Durability and freedom from maintenance

On exposure to air and moisture, unprotected steel corrodes to form an oxide coating, i.e. rust, which is permeable to moisture and thus encourages progressive corrosion which may in time adversely affect the strength of the material. To inhibit rust, steel is either painted or coated with zinc. Painted surfaces require periodic repainting. Zinc coatings that are perforated by cutting and drilling will not protect the exposed steel below which will corrode progressively.

Concrete which is solidly compacted around steel reinforcement will provide very good protection against corrosion of the steel reinforcement. Hair cracks that may form in concrete encasement to steel, due to the drying shrinkage of concrete, will allow moisture to penetrate to the steel reinforcement which will corrode. The rust formed by corrosion expands and will in time cause the concrete cover to spall away from steel and encourage progressive corrosion.

As protection against corrosion the reinforcement may be zinc coated or stainless steel reinforcement may be used. Zinc coating adds little to cost whereas stainless steel adds appreciably. Thoroughly mixed, solidly consolidated concrete will have a dense, smooth surface that requires no maintenance other than occasional steam cleaning.

Adequately seasoned (dried), stress-graded timber preserved against fungal and insect attack should require no maintenance during the useful life of a building other than periodic staining or painting for appearance.

Fire safety

The requirements for the fire resistance of structure in the Building Regulations 1991 do not apply to roof structures unless the roof is used as a floor, nor to single-storey structures supporting a roof except where a wall is close to a boundary and is required to have resistance to the spread of fire between adjacent buildings having regard to the height, use and position of the building.

TRUSS CONSTRUCTION

...russ', in connection with roof frames, is used in the sense of defining the action of a triangular roof framework where the spread under load of sloping rafters is resisted by the horizontal tie member, secured to the feet of the rafters, which trusses or ties them against spreading. 'Lattice' is used in the sense of an open grid or mesh of slender members fixed across or between each other, generally in some regular pattern of cross-diagonals or as a rectilinear grid.

Symmetrical pitch steel lattice truss construction

The simple, single bay shed frame illustrated in Fig. 2 is to this day one of the cheapest forms of structure. The small section, mild steel members of the truss can be cut and drilled with simple tools, assembled with bolted connections and speedily erected without the need for heavy lifting equipment. The small section, steel angle members of the truss are bolted to gusset plates. The end plates of trusses are bolted to columns and purlins, and sheeting rails are bolted to cleats bolted to rafters and columns respectively, to support roof and side wall sheeting. The considerable depth of the roof frames at mid span provides adequate strength in supporting dead and imposed loads and rigidity to minimise deflection under load. The structural frames and their covering provide basic shelter for a variety of uses.

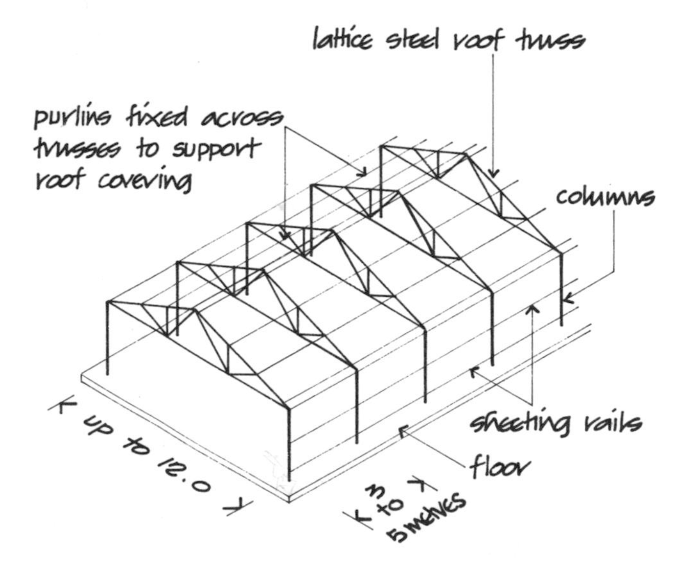

Fig. 2

The advantage of this simple, single-storey, single-bay frame is economy in the use of materials by the use of small section angle, tubular or flat standard mild steel sections for the trusses that can be economically fabricated and quickly erected on comparatively slender mild steel I-section columns fixed to concrete pad foundations.

The disadvantages of this structural framework are the very considerable volume of roof space inside the triangular roof frames that cannot be used for any purpose other than housing services such as lighting and heating, and the considerable visible area of roof that is not generally accepted as an attractive feature of small single-storey buildings. Where the activity enclosed by the building requires heating, the roof space has to be wastefully heated as well as the useful space below. For maximum structural efficiency the slope or pitch of the rafters of these frames should be not less than about 17° to the horizontal. For economy in the use of small-section framing for the trusses and to limit the volume of unused roof space, these trusses are generally limited to spans of 12.0 and for economy in the use of small section purlins and sheeting rails the spacing of trusses is usually betwen 3.0 to 5.0.

The bolted, fixed base connection of the foot of the columns to the concrete foundation bases provides sufficient strength and stability against wind pressure on the side walls and roof. Wind bracing provides stability against wind pressure on the end walls and gable ends of the roof.

Because of the limited penetration of daylight through side wall windows, a part of the roof is often covered with glass or translucent plastic sheets which are fixed in the slope of roofs, usually in the middle third of each slope as illustrated in Fig. 3, to provide reasonable penetration of daylight to the working surfaces in the building.

The thin sheets of profiled steel sheet that are commonly used to provide cover to the walls of these buildings have poor resistance to damage by knocks. As an alternative to steel columns to support the roof trusses, brick side walls may be used for single-bay buildings to provide support for the roof frames, protection against wind and rain and solid resistance to damage by knocks. Figure 4 is an illustration of a single-bay single-storey building with brick side walls supporting steel roof trusses. The side walls are stiffened by piers formed in them under the roof trusses. As an alternative, a low brick upstand wall may be raised either outside of or between the columns as protection against knocks, with wall sheeting above.

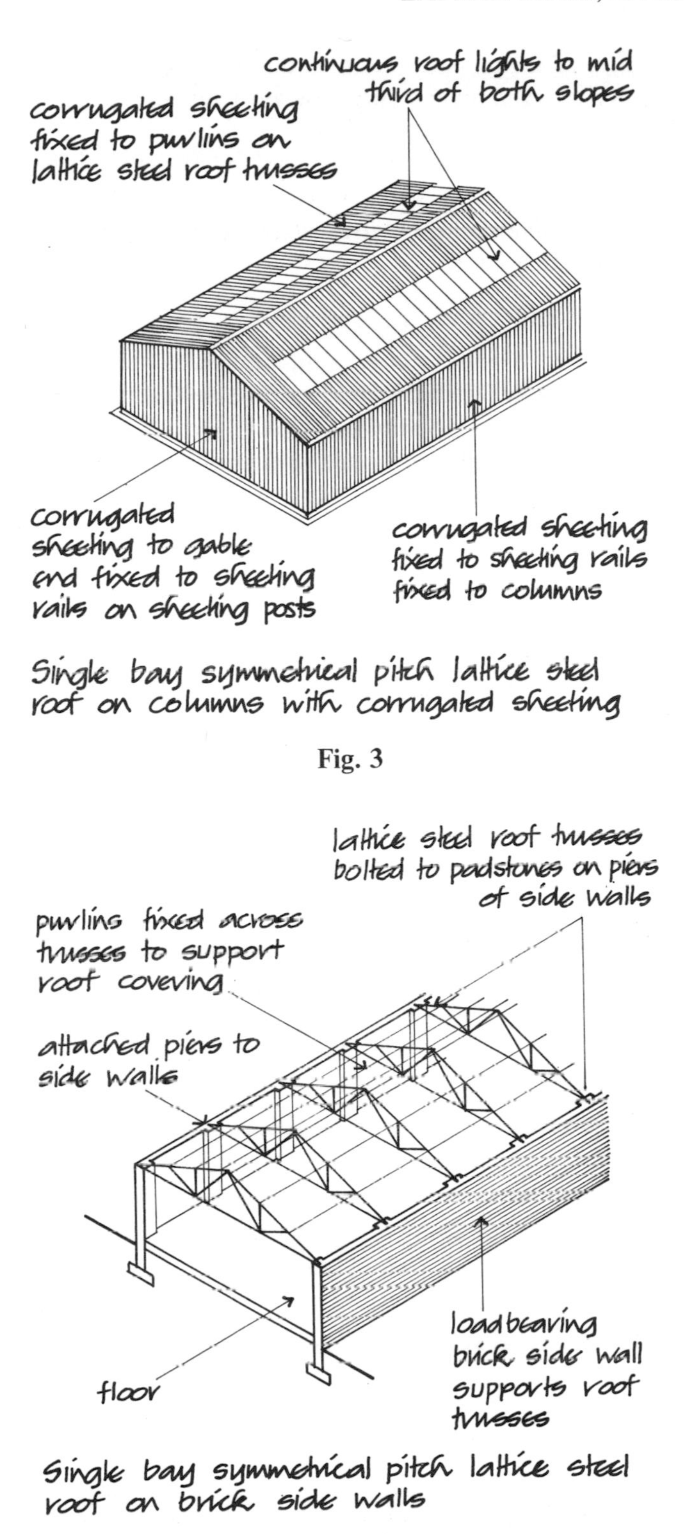

Single bay symmetrical pitch lattice steel roof on columns with corrugated sheeting

Fig. 3

Single bay symmetrical pitch lattice steel roof on brick side walls

Fig. 4

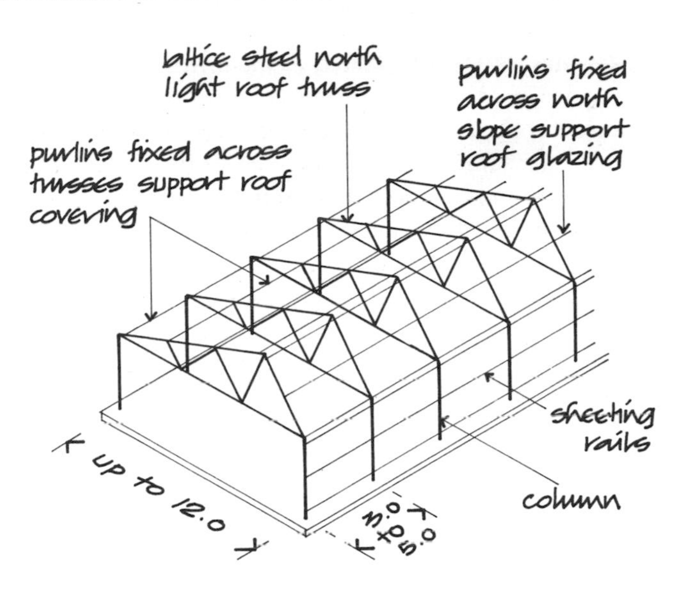

Single bay north light lattice steel roof trusses on steel columns

Fig. 5

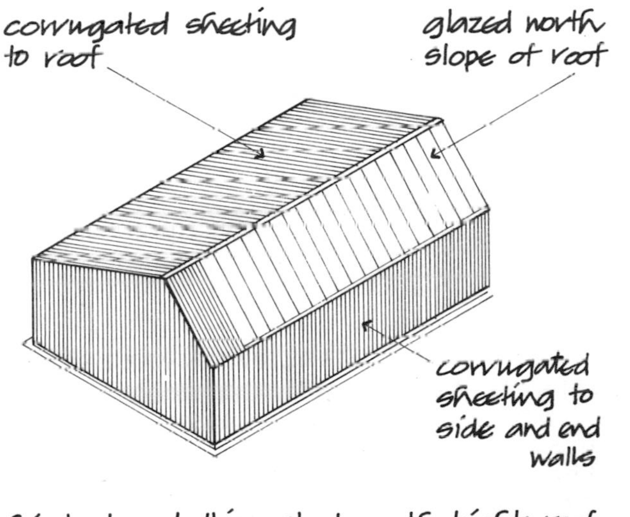

Single bay lattice steel north light roof on columns with corrugated sheeting

Fig. 6

North light steel lattice truss construction

Rooflights in the slopes of symmetrical pitch roofs, which are generally set in east and west facing slopes, may cause discomfort through overheating in summer and disrupt manufacturing activities by the glare from sunlight. To avoid these possibilities the north light roof is used. The north light roof has an asymmetrical profile with the south facing slope at 17° or more to horizontal and the north facing slope at from 60° to vertical. Figure 5 is an illustration of a single-bay, single-storey building with north light, steel lattice trusses on columns. The whole of the south slope is covered with profiled sheets and the whole of the north facing slope with glass or clear or translucent plastic sheeting as illustrated in Fig. 6.

Because of the steep pitch of the north facing slope the space inside the roof trusses of a north light roof is considerably greater than that of a symmetrical pitch roof of the same span as illustrated in Fig. 7. To limit the volume of roof space that cannot be used and has to be wastefully heated, most north light roofs are limited to spans of up to about 10.0.

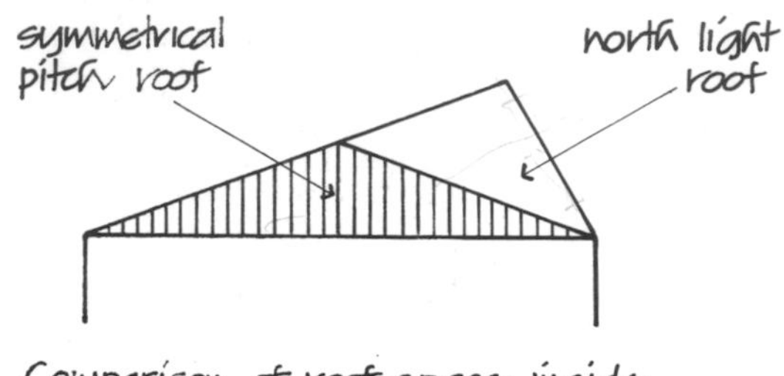

Comparison of roof space inside symmetrical pitch and north light roofs of the same span

Fig. 7

Multi-bay lattice steel roof truss construction

There is no theoretical limit to the span of a single-bay, steel roof truss to provide clear unobstructed floor area. For structural efficiency a triangular truss should have a pitch of not less than 17° to the horizontal. With increase in span there is an increase in the volume of unused space inside the roof trusses and the length of truss members. To cover large areas it is, therefore, usual to use two or more bays of symmetrical pitch roofs to limit the volume of roof space and length of members of the trusses. Figure 8 is an illustration of the comparative volume of a single long span roof and four smaller roof bays covering the same floor area.

To avoid the use of closely spaced internal columns to support roof trusses it is usual with multi-bay roofs to use either valley beams or lattice girders inside the depth of the trusses to reduce the number of internal columns that would otherwise obstruct the working floor area.

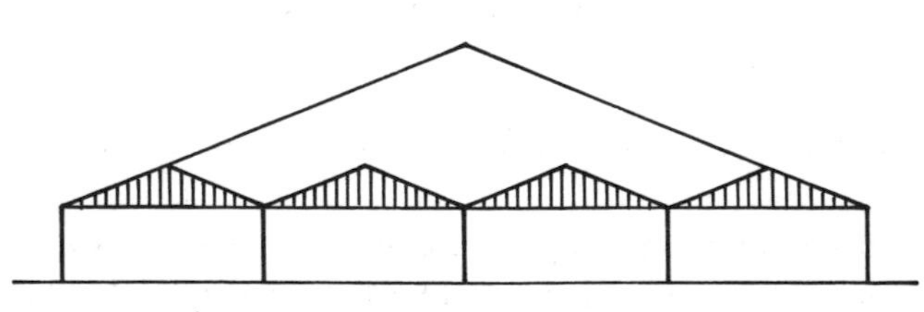

Comparison of volume of roof space and area of truss of one single and four trusses

Fig. 8

Multi-bay valley beam lattice steel roof truss construction

A beam under the valley of adjacent roofs supports the ends of roof trusses between the internal columns that support the valley beam as illustrated in Fig. 9. Plainly the greater the span or space between internal columns supporting a valley beam the greater will be the depth of the valley beam, so that for a given required clear working height, an increase in the depth of the valley beam will increase the volume of unused roof space above the underside or soffit of the valley beam as illustrated in Fig. 10.

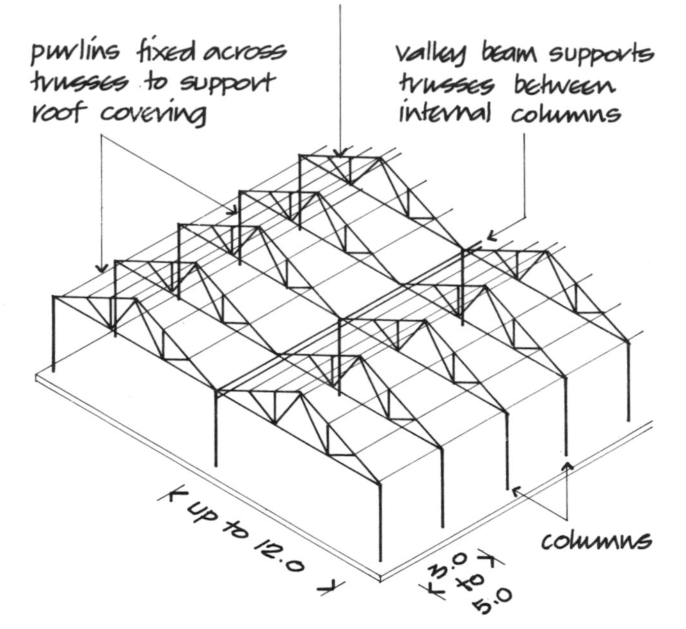

Two bay symmetrical pitch lattice steel roof and columns with valley beam

Fig. 9

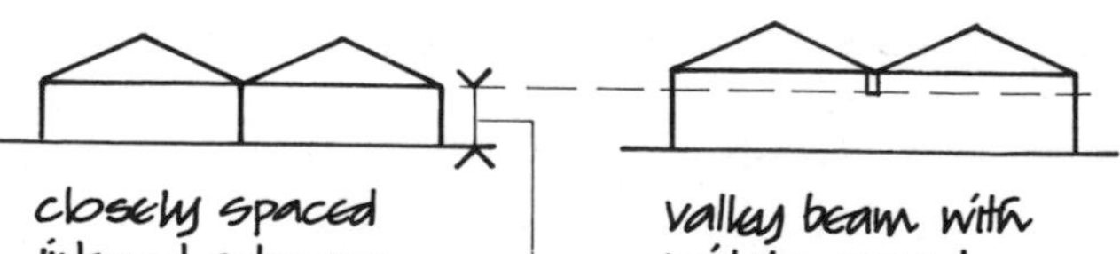

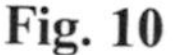

Fig. 10

Similarly a valley beam may be used in multi-bay north light truss roofs as illustrated in Fig. 11.

A disadvantage of the multi-bay valley beam form of construction is that there is very limited depth alongside the valley beam for the fall (slope) of rainwater pipes from valley gutter outlets to rainwater down pipes fixed to internal columns. The shallow fall rainwater pipes that are run alongside valley beams will require sealed joints and the shallow fall pipe will more readily become blocked than a straight down pipe from valley outlets.

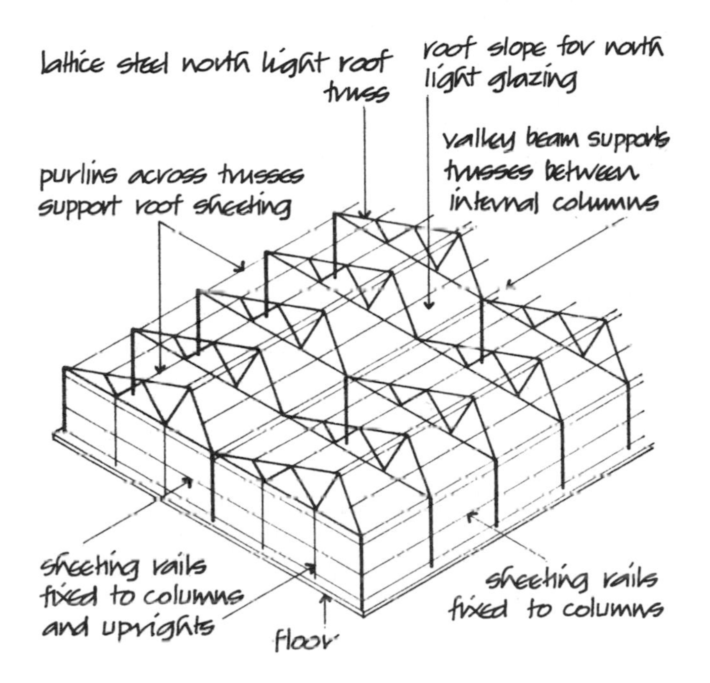

Two bay north light lattice steel roof with columns and valley beam

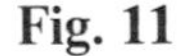
Fig. 11

Cantilever (umbrella) multi-bay lattice steel truss roof construction

A lattice girder constructed inside the depth of each bay of symmetrical pitch roof trusses will, because of its great depth, be capable of supporting the roof between comparatively widely spaced internal columns and will not project below the underside of trusses. Figure 12 is an illustration of a cantilever or umbrella roof with lattice steel girders constructed inside the depth of each bay of trusses at mid span. The lattice girder supports half of each truss with each half cantilevered each side

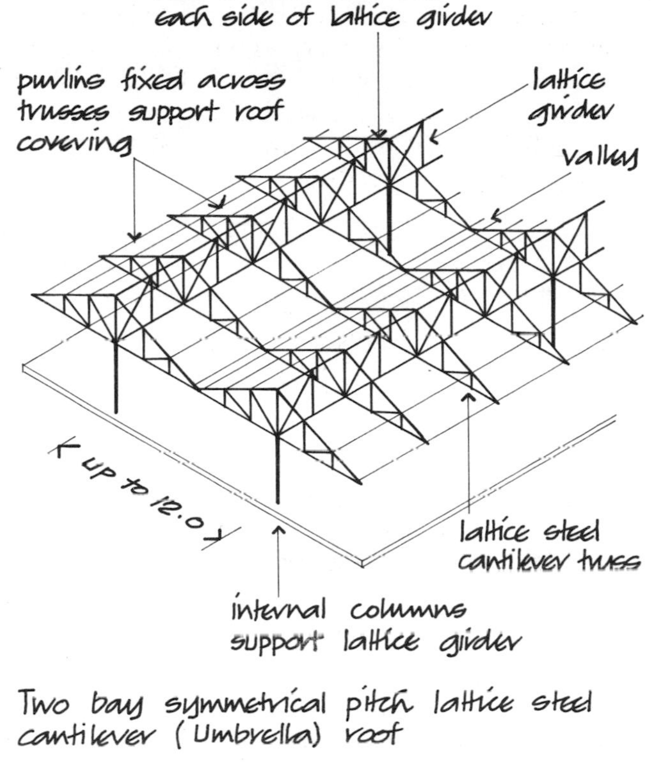

Two bay symmetrical pitch lattice steel cantilever (Umbrella) roof

Fig. 12

of the truss, hence the name 'cantilever truss' roof. The outline in section of the column and the truss cantilevered each side of the lattice girder resembles an umbrella, hence the name 'umbrella' roof.

The principal disadvantage of the umbrella roof form of construction is that the comparatively widely spaced internal columns provide very little fall for rainwater pipes from valley gutter outlets to rainwater down pipes fixed to internal columns under the lattice girders inside roof frames. To provide an appreciable fall to rainwater pipes a modified form of umbrella roof is used with the tie or chord member of trusses framed on the slope as illustrated in Fig. 13 to provide some fall for rainwater pipes draining from valley gutters.

North light multi-bay lattice steel truss construction

North light trusses may be supported by a lattice girder, as illustrated in Fig. 14, with widely spaced internal columns to cover large areas with the least obstruction. The profile of a multi-bay north light roof resembles the teeth of a saw, hence the name 'saw-tooth' roof that is often used for this type of roof.

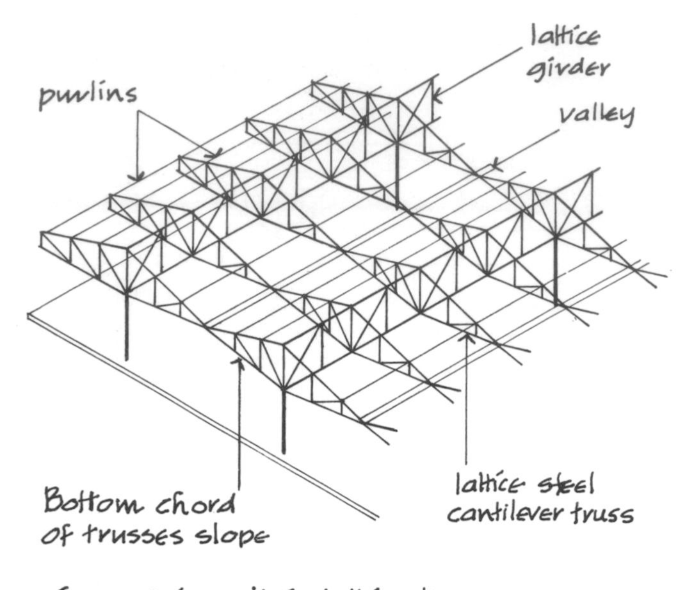

Fig. 13

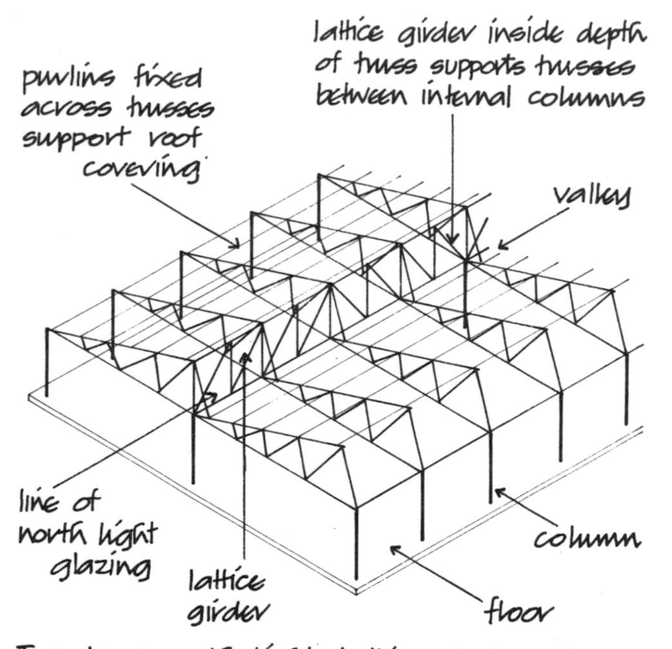

Fig. 14

The disadvantage of the cantilver roof form is the great number of lattice members in the roof of both trusses and the lattice girders as these will collect dust and dirt, need frequent painting to inhibit rust and will, to some extent, obstruct natural light from roof lights.

Lattice steel truss construction

For the sake of economy in using one standard section, lattice steel trusses are often fabricated from one standard steel angle section with two angles, back to back, for the rafters and main tie and a single angle for the internal struts and ties as illustrated in Fig. 15.

The usual method of joining the members of a steel truss is by the use of steel gusset plates that are cut to shape to contain the required number of bolts at each connection. The flat steel gusset plates are fixed between the two angle sections of the rafters and main tie and to the intermediate ties and struts, as illustrated in Fig. 15. Bearing plates fixed to the foot of each truss provide a fixing to the cap of columns. The members of the truss are bolted together through the gusset plates.

Standard I-section steel columns are used to support the roof trusses. A steel base plate is welded or fixed with bolted connections with gusset plates and angle cleats to the base of the columns. The column base plate is levelled on a grout of cement on the concrete pad foundation to which it is rigidly fixed with four holding down bolts, cast or set into the foundation, as illustrated in Fig. 16. The rigid fixing of the columns to the foundation bases provides stability to the columns which act as vertical cantilevers in resisting lateral wind pressure on the side walls and the roof of the building. A cap is welded or fixed with bolted connections to the top of each column and the bearing plates of truss ends are bolted to the cap plate as illustrated in Fig. 16.

Lattice trusses can be fabricated from tubular steel sections that are cut, mitred and welded together as illustrated in Fig. 17. Because of the labour involved in cutting and welding the members, a tubular steel section truss is more expensive than an angle section truss. From the economy of fabricating standard trusses and the economy of repetition in producing many similar trusses a tubular section truss may be only a little more expensive than a similar one-off angle section truss.

The advantages of the tubular section truss are the greater structural efficiency of the tubular section over the angle section and the comparatively clean line of the tubulars and their welded connections which reduce the surface area liable to collect dust and requiring paint. The truss illustrated in Fig. 17 has a raised tie, the middle third of the length of the main tie being raised above the level of the foot of the trusses. This raised tie affords some increase in working height below the raised part of the tie which plainly is only of advantage with medium and long span roofs.

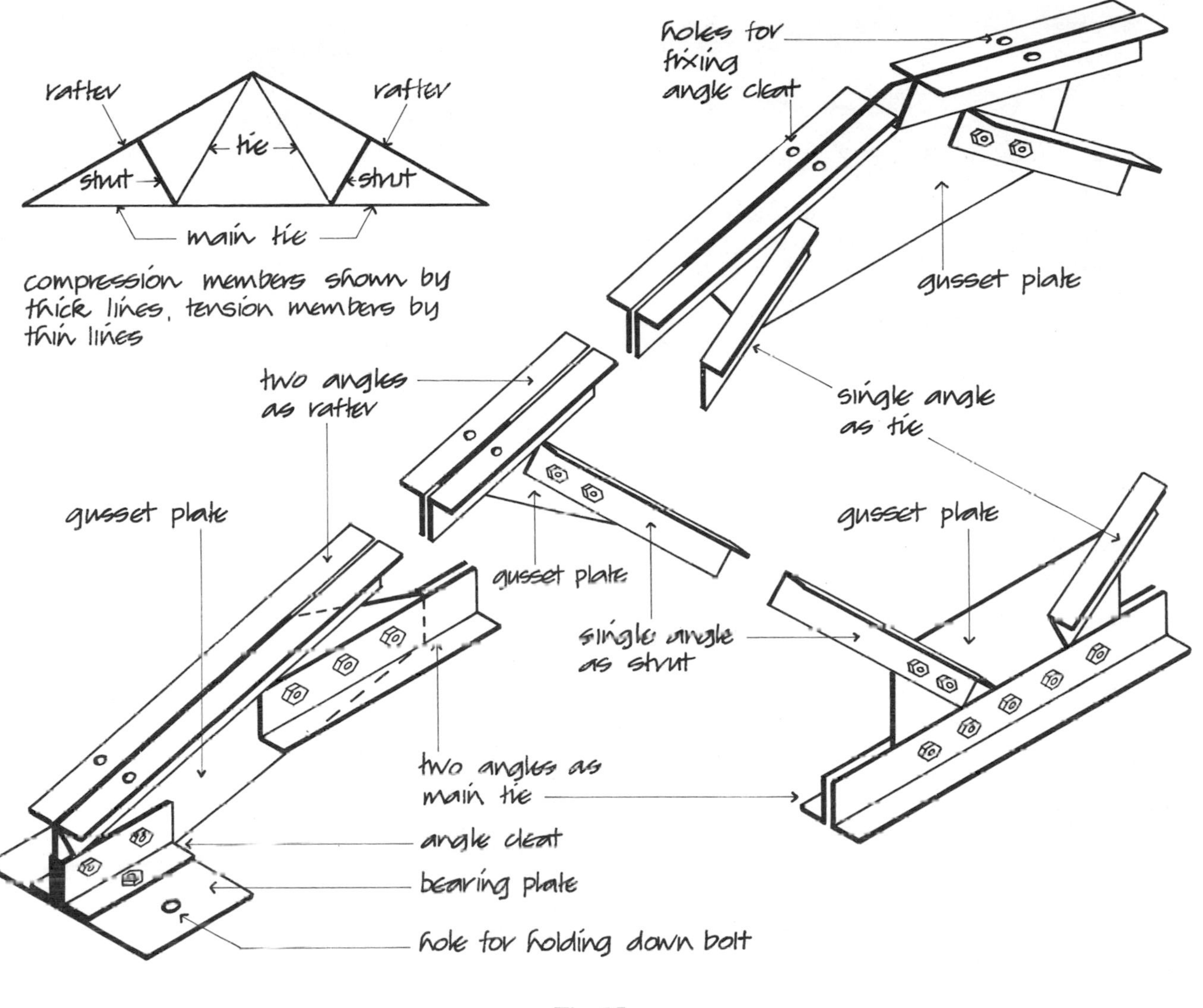

Fig. 15

STEEL LATTICE BEAM ROOF CONSTRUCTION

The introduction of deep profiled steel sheeting, which can be laid as a roof covering to roofs pitched as low as 6° to the horizontal, coincided with the general adoption of space heating for many single-storey buildings and regulations requiring insulation to conserve fuel. In consequence the traditional lattice truss form of roof was no longer the most economical or suitable form of roof for buildings that required heating because of the considerable volume of unusable roof space and the requirement of a pitch of at least 17° for the structural efficiency of a truss roof. The two structural forms best suited to the use of deep profiled steel roof sheeting are lattice beam and portal frame forms of construction.

The simplest form of lattice beam roof is a single-bay symmetrical pitch roof constructed as a cranked lattice beam or rafter as illustrated in Fig. 18. Lattice beam roof frames are fabricated as uniform depth, symmetrical pitch cranked (bent) beams, uniform depth mono-pitch roofs, tapering depth butterrfly roofs and tapering depth beams.

Symmetrical pitch lattice steel beam roof construction

The uniform depth lattice beam is cranked to form a symmetrical pitch roof with slopes of from 5° to 10° to the horizontal as illustrated in Fig. 18. The beams are generally fabricated from tubular and hollow rectangu-

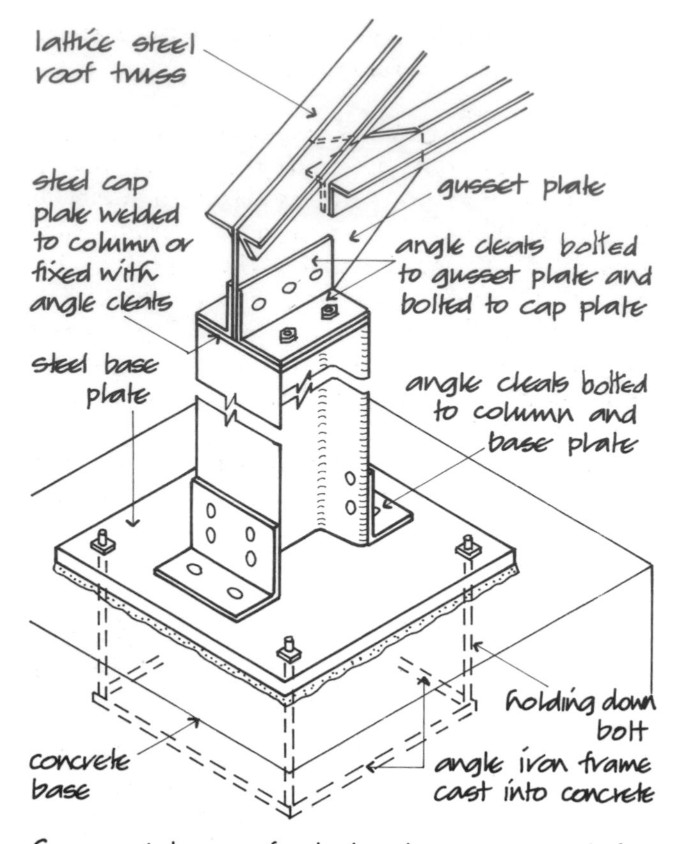

Fig. 16

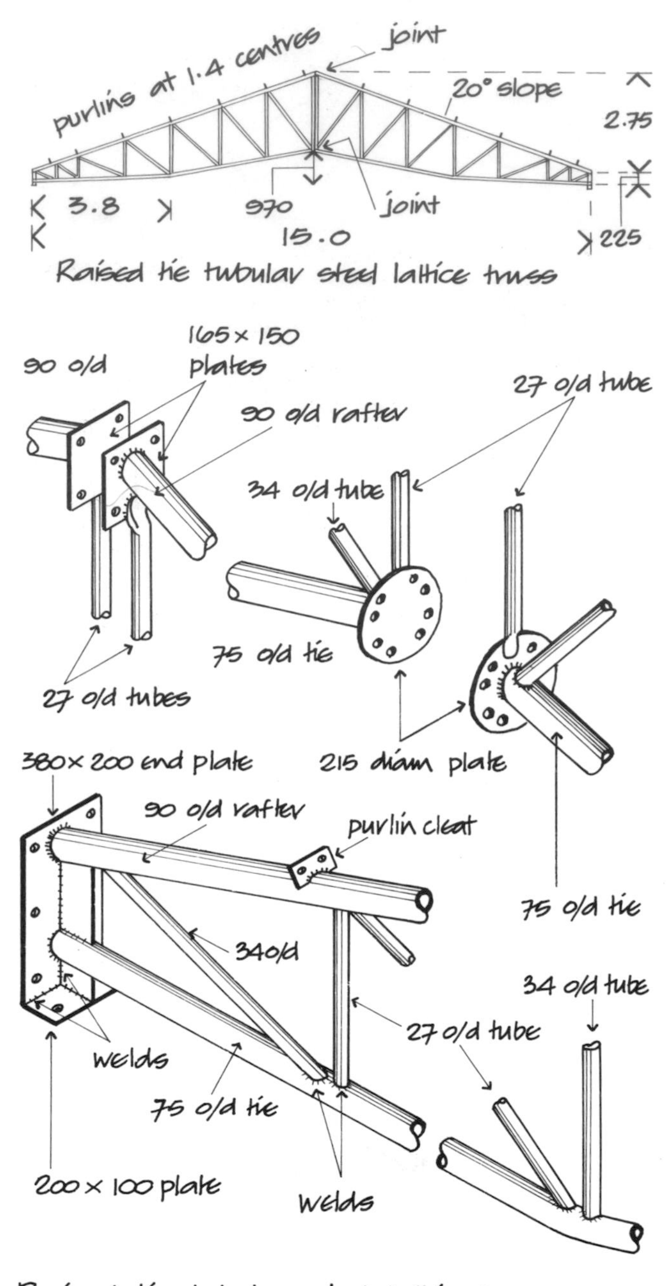

Fig. 17

lar section steel sections that are cut and welded together with bolted site connections at mid span to facilitate transport of half lengths. For convenience in making straight, oblique cut ends to the intermediate tubulars of the lattice members, the top and bottom chord of the beams are usually of hollow rectangular section. End plates welded to the lattice beams are bolted to the flanges of I-section columns. With lattice beam roof frames, service pipes and small ducts may be run through the lattice frames and larger ducts slung below the beams inside the unused roof space. Because of the low pitch or slope of this roof form there is little unused roof space inside the shallow lattice beams and below the beams.

Multi-bay symmetrical pitch valley beam lattice beam roof construction

Because of the shallow depth of the lattice beam it is not practical to construct an umbrella type of roof with a deep beam inside the depth of frames below the ridge. For multi-bay symmetrical pitch lattice beam roofs it is usual to fabricate a form of valley beam roof as illustrated in Fig. 19. So that there is no increase in unused roof space the valley beam has to be of the same shallow depth as the beams and in consequence may only provide support for every other internal system of roof frames and thus give only a small increase in free floor space. The shallow depth of the valley beams provides minimum fall for rainwater pipes run from valley gutter outlets to down pipes fixed to internal columns.

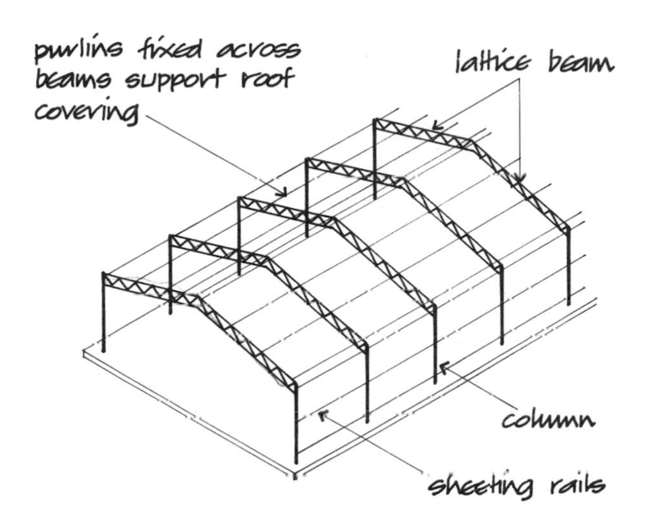

Fig. 18

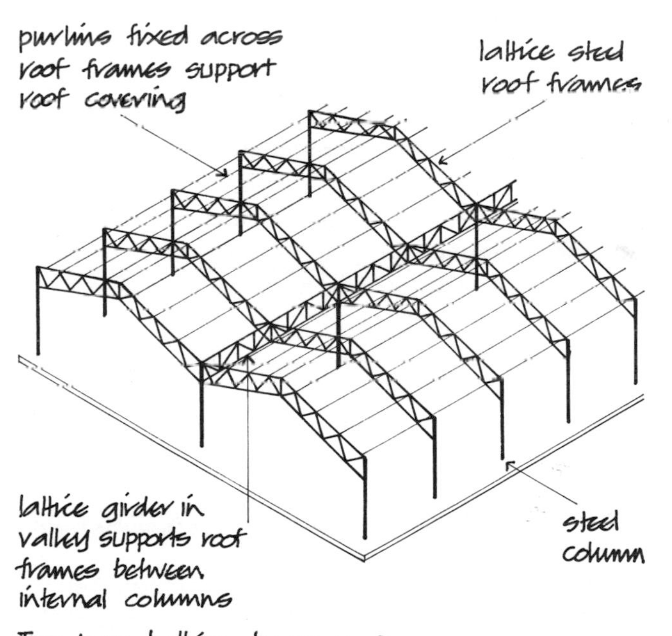

Fig. 19

Multi-bay symmetrical pitch valley beam lattice steel beam (butterfly) roof construction

To provide the maximum free floor area a form of butterfly roof, with deep valley beams supporting tapered lattice beams, is used as illustrated in Fig. 20. The lattice beams taper from the supporting valley beams up to the ridge in each bay. The deeper the valley beams the greater the spacing between internal columns and the greater the unused roof space inside and below the frames. The depth of the valley beams provides adequate depth for the fall of rainwater pipes run to down pipes.

The name of this roof form, butterfly, derives from the shape of the tapered rafters that resembles the wings of a butterfly.

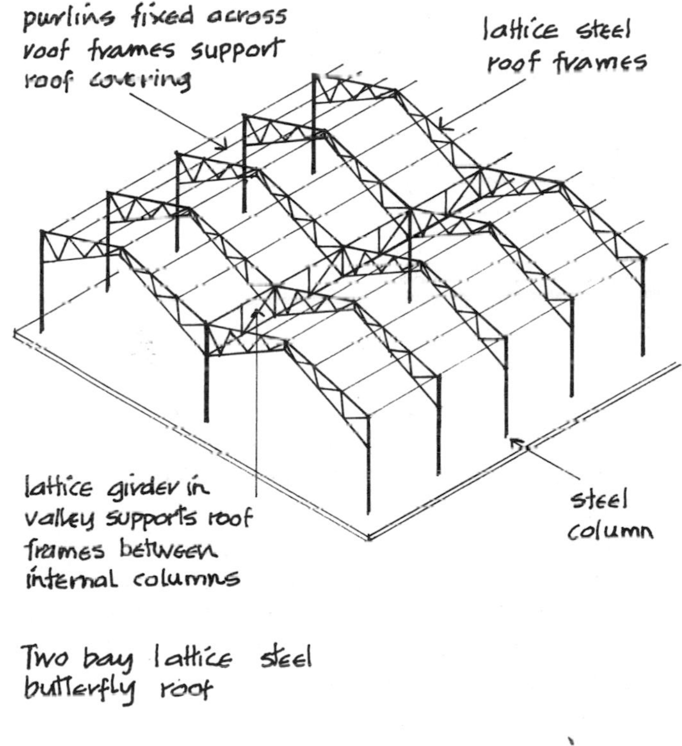

Fig. 20

Multi-bay monopitch north light (sawtooth) lattice steel beam roof construction

Because the depth of a lattice beam roof frame minimises deflection under load, it can be used as a monopitch form of construction for low or very low pitch roofs, either as a single bay roof or as a multi-bay roof.

The two bay monopitch roof illustrated in Fig. 21 is fabricated as a north light form of roof with a deep valley beam at the junction of adjacent roofs to allow wide spacing of columns. The depth of the lattice beam depends on the span and slope of the adjacent monopitch roofs, the greater the span and slope of the roofs, the deeper the valley beam, the wider the spacing of internal columns and the greater the unused roof space.

The lattice beams shown in Fig. 21 taper towards the eaves to reduce, to some extent, the volume of unused roof space and the braces from the roof frames to columns provide some stiffening against overturning. There is very little depth below valley gutters in this form of roof for an adequate fall of rainwater pipes run from valley gutter outlets to down pipes fixed to internal columns.

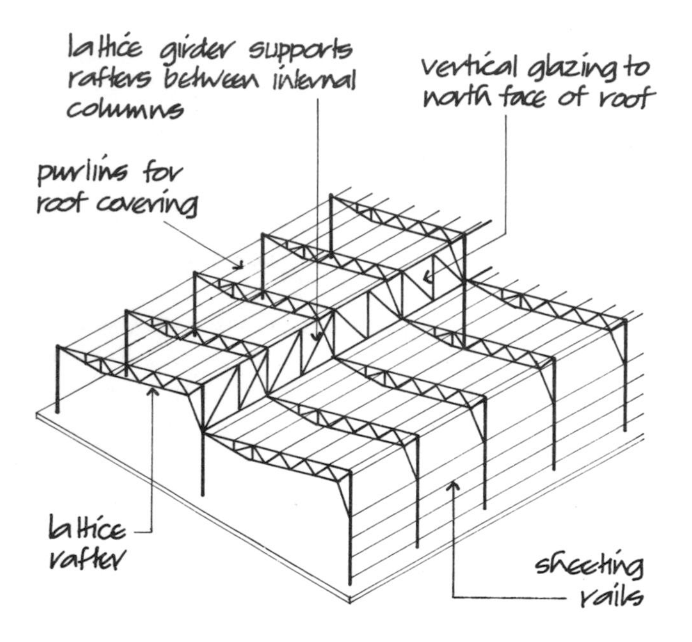

Fig. 21

PORTAL FRAMES

Steel portal frames

Following the acceptance of the plastic theory of design, proposed by Professor Baker, rigid portal frames became an economic alternative to lattice truss and lattice beam roofs.

The application of the plastic theory in place of the previous, generally used, elastic method of design (see Volume 4) is particularly relevant to rigid frames of a ductile material such as steel. The plastic theory takes account of the distribution of moments through the whole of the rigid frame under working loads so that sections lighter and more slender than sections determined by the elastic method of design may be used safely.

To be effective a pitched roof portal frame should have as low a pitch as practical to minimise spread at the knee of the portal frame (spread increases with the pitch of the rafters of a portal frame). The knee of a portal frame is the rigid connection of the rafter to the post of the portal.

The early use of the rigid portal frame coincided with the introduction of a wide range of cold formed, profiled steel sheets for roofing, which could be fixed at a low pitch and be weather-tight. The combination of low pitch steel portal frames and profiled steel roof sheeting and decking has led to the adoption of this form of structure, particularly for single-bay single-storey buildings.

A portal frame is distinguished by the rigid connection of the rafters to the posts of the frame so that under load moments are distributed through the rafter and the post. For short- and medium-span frames the apex or ridge, where the rafters connect, is generally made as an on-site, rigid bolted connection for convenience in transporting half portal frames. Long-span portal frames may have a pin joint connection at the ridge to allow some flexure between the rafters of the frame which are pin jointed to foundation bases to allow flexure of posts due to spread at the knees under load.

For economy in the use of a standard section, short- and medium-span steel portal frames are often fabricated from one mild steel I-section for both rafters and posts, with the rafters welded to the posts without any increase in depth at the knee as illustrated in Fig. 22.

Short-span portal frames may be fabricated off site as one frame. Medium-span portal frames are generally fabricated in two halves for ease of transport and are assembled on site with bolted connections of the rafters at the ridge, with high strength friction grip (hsfg) bolts (see Volume 4).

Many medium- and long-span steel portal frames have the connection of the rafters to the posts at the knee, haunched to make the connection deeper than the main rafter section for additional stiffness as illustrated in Fig. 23. In long-span steel portal frames the posts and lowest length of the rafters, towards the knee, may

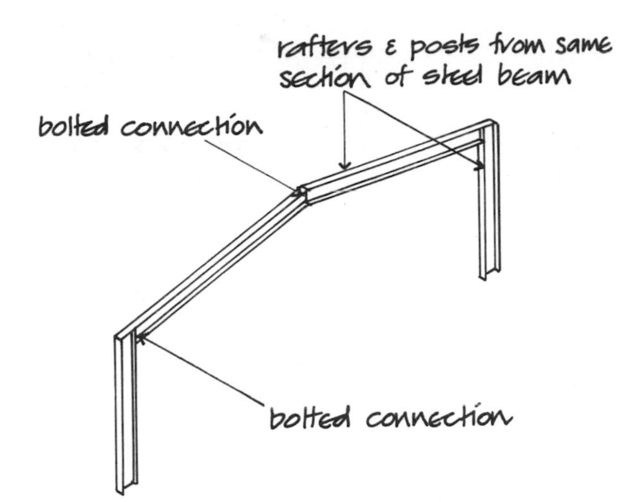

Short span steel portal frame

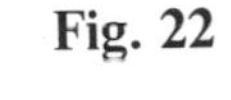

Fig. 22

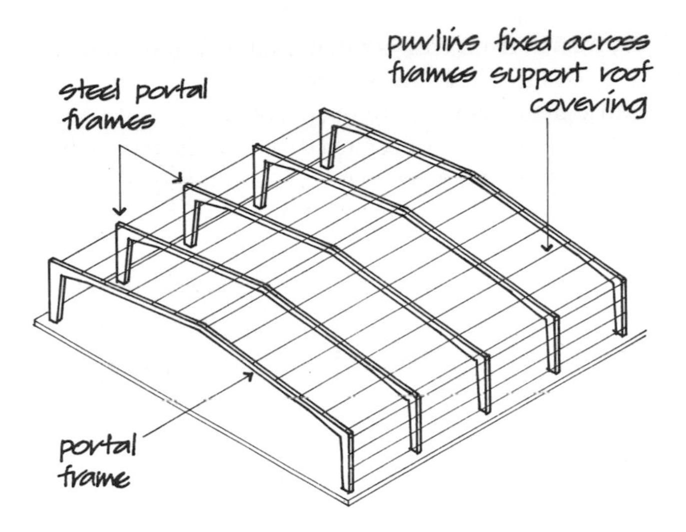

Long span steel portal frames

Fig. 24

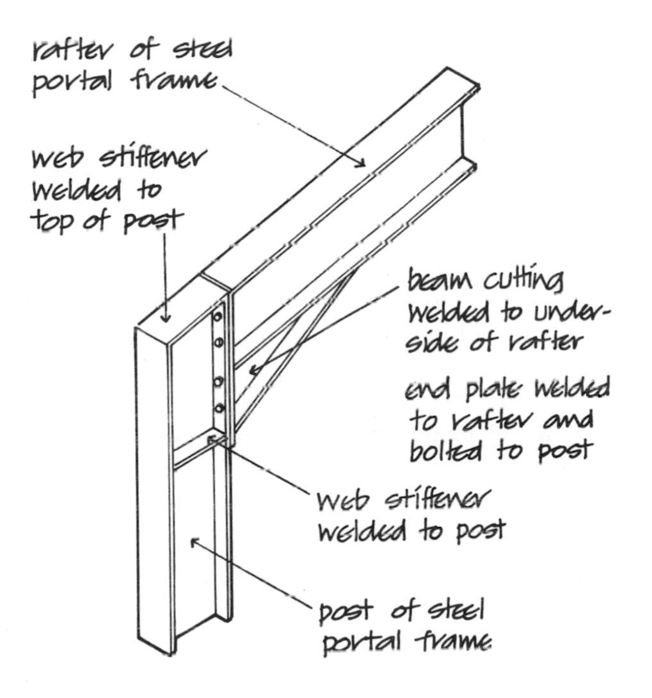

Haunch to steel portal frame

Fig. 23

often be fabricated from cut and welded I-sections so that the post section and part of the rafter is wider at the knee than at the base and ridge of the rafter (Fig. 24).

The haunched connection of the rafters to the posts can be fabricated either by welding a cut I-section to the underside of the rafter, as illustrated in Fig. 23, or by cutting and bending the bottom flange of the rafter and welding in a steel gusset plate.

The junction of the rafters at the ridge is often stiffened by welding cut I-sections to the underside of the rafters at the bolted site connection as shown in Fig. 25.

Steel portal frames may be fixed to or pinned to bases to foundations. For short-span portal frames, where there is comparatively little spread at the knee or

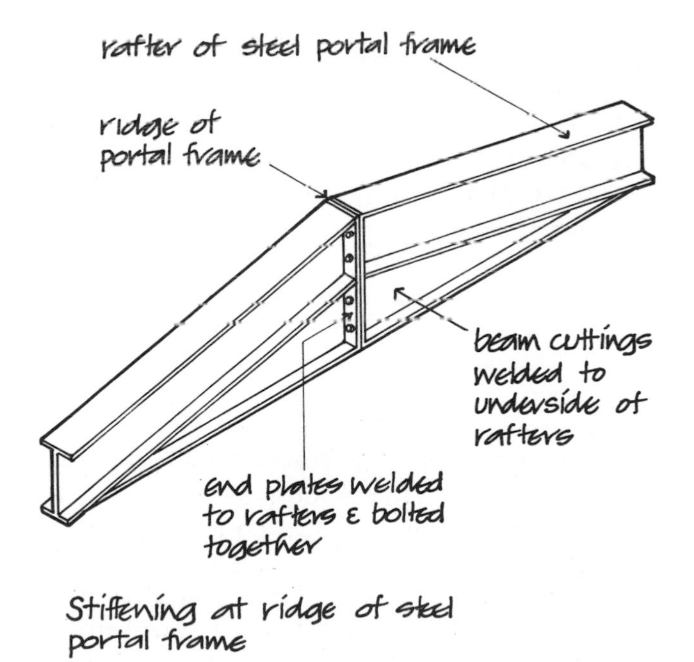

Stiffening at ridge of steel portal frame

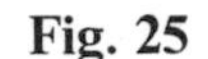

Fig. 25

haunch, a fixed base is often used. It will be seen from Fig. 26 that the steel base plate, which is welded through gusset plates to the post of the portal frame, is set level on a bed of cement grout on the concrete pad foundation and is secured by four holding-down bolts set or cast into the concrete foundation.

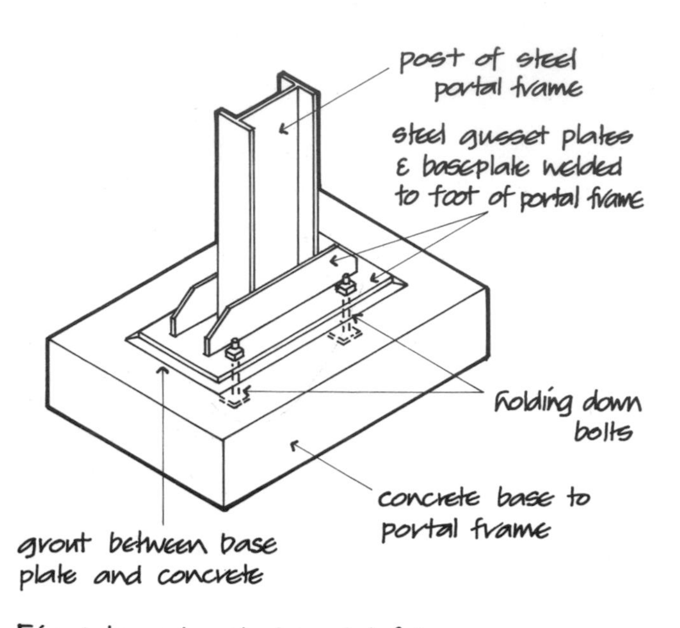

Fig. 26

A pinned base is made by sitting the portal base plate on a small steel packing on to a separate base plate bearing on the concrete foundation. Two anchor bolts, either cast or set into the concrete pad foundation, act as holding-down bolts to the foot of the portal frame as illustrated in Fig. 27. This type of base is described as a pinned base as the small packing between the two plates allows some flexure of the portal post independent of the foundation which in consequence may be less substantial than a comparable fixed base.

Portal frames with a span of up to 15.0 are defined as short span, frames with a span of 16.0 to 35.0 as medium span and frames with a span of 36.0 to 60.0 as long span.

Short-span portal frames are usually spaced at from 3.0 to 5.0 apart and medium-span portal frames at from 4.0 to 8.0 apart to suit the use of angle or cold formed purlins and sheeting rails. Long-span steel portal frames are usually spaced at from 8.0 to 12.0 apart to economise in the number of comparatively expensive frames, with channel, I-section or lattice purlins and

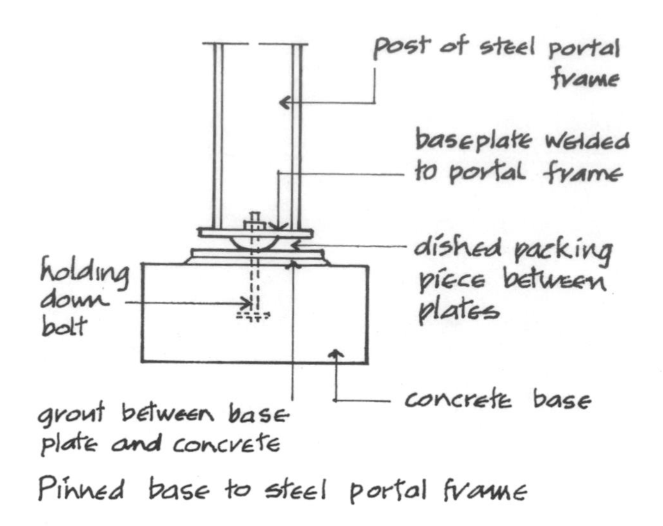

Fig. 27

sheeting rails to support roof sheeting or decking and walling.

With flat or low pitch steel portal frames it is difficult to ensure a watertight system of roof glazing fixed in the slope of the roof, with either glass or plastic sheets. For natural roof lighting a system of monitor roof lights is sometimes used. These lights are formed by welded, cranked I-section steel purlins fixed across the portal frames as illustrated in Fig. 28. The monitor lights project above the roof with two upstand faces that may be vertical or sloping. The monitor lights shown in Fig. 28 are designed with the vertical faces facing south to minimise the direct penetration of sunlight and the sloping faces facing north to provide a good distribution of natural light to the interior. The monitor lights can be constructed to provide natural or controlled ventilation. The monitor lights finish short of the eaves to avoid the difficulty of the finish at eaves that would otherwise occur.

Because of the very considerable spans practical with steel portal frames there is little if any advantage in the use of multi-bay steel portal systems.

Bracing

Wind bracing The side wall columns (stanchions) and their fixed bases that support the roof frames of a single-bay, single-storey structure are designed to act as vertical cantilevers to carry the loads in bending and shear that act on them from horizontal wind pressure on the sheeting fixed to purlins and roof frames and

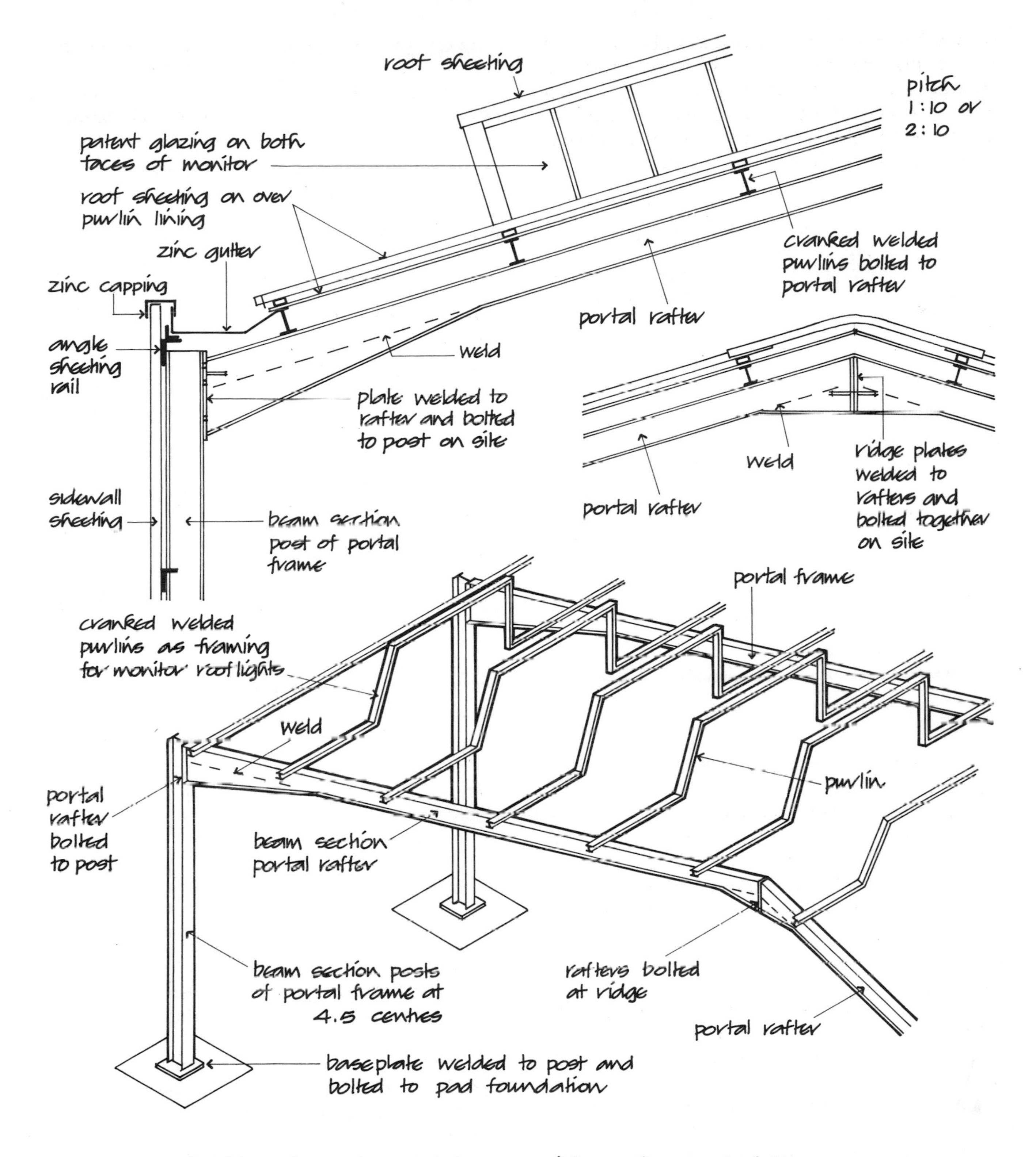

Solid web steel portal frame with monitor roof lights

Fig. 28

side wall sheeting fixed to sheeting rails. The rigid knee joint of rafters to posts of portal frames is generally sufficient to carry the loads from horizontal wind pressure on roof and side wall sheeting.

Where internal columns to multi-bay, single-storey buildings are comparatively widely spaced under roof lattice girders or beams supporting intermediate roof frames of valley, umbrella, butterfly, north light and saw tooth roof systems, it is generally necessary to use a system of eaves bracing to assist in the distribution of horizontal loads from wind pressure on side walls and roof, between the external and the more widely spaced internal columns. The system of eaves bracing shown in Fig. 29 consists of steel sections fixed between the tie or bottom chord of roof frames and columns.

between roof frames serve to stabilise the frames against probable uplift due to wind pressure.

The vertical bracing in the adjacent wall frames at gable end corners assists in setting out and squaring up the building and also serves as bracing against wind pressure on the gable ends of the building.

Purlins and sheeting rails

Purlins are fixed across rafters and sheeting rails across the columns of single-storey frames to provide support and fixing for roof and wall sheeting and insulation. The spacing of the purlins and sheeting rails depends on the type of roof and wall sheeting used. The deeper the

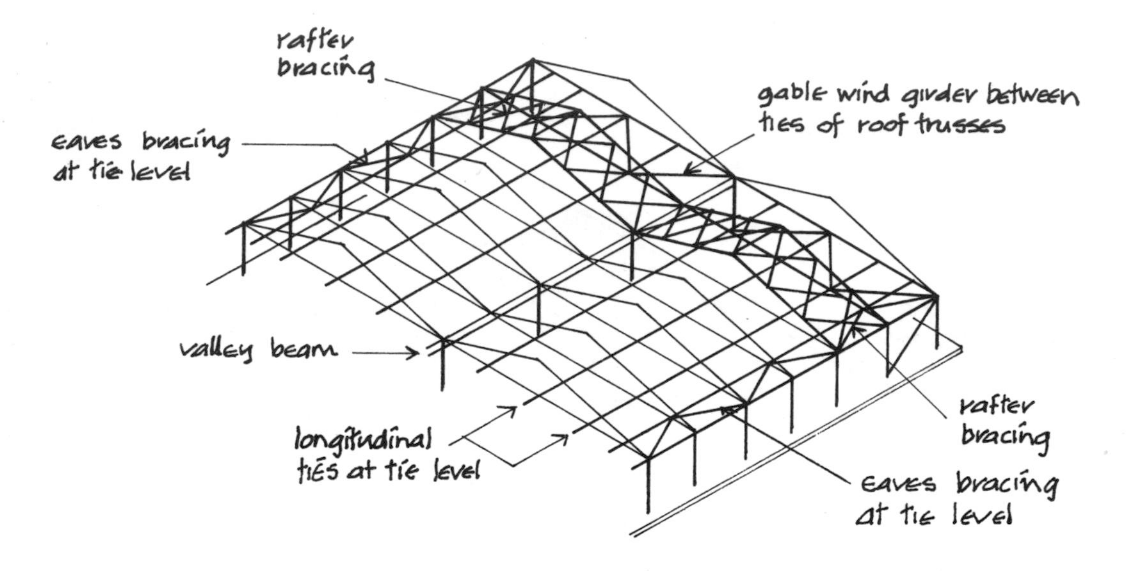

Fig. 29

profile of sheeting the greater its safe span and the further apart the purlins and sheeting rails may be fixed.

To transfer the loads from wind pressure on the end wall sheeting and vertical sheeting rails of gable ends, a system of horizontal gable girders is formed at tie or bottom chord level as illustrated in Fig. 29.

Structural bracing Other bracing to the frames is used to assist in setting out the building, to stabilise the roof frames and square up the ends of the building.

The rafter bracing between the end frames, illustrated in Fig. 29 serves to assist in setting out the building frames and to stabilise the rafters of the roof frames. Longitudinal ties, illustrated in Fig. 29,

The section of the purlins and sheeting rails depends on the most economic spacing of the structural frames. The greater the spacing of frames the greater the dead-weight of sheeting and imposed loads, and the deeper the section of purlin and rail necessary to support the weight of the roof and wall covering and loads from wind and snow.

Before 1960 most purlins and sheeting rails were of standard mild steel sections, angle sections being common for closely spaced frames and channel sections for more widely spaced frames. Angle and channel sections were suited to the hook bolt fixings then used for corrugated asbestos cement and steel sheeting.

Angle and channel section purlins and sheeting rails are fixed to short lengths of steel angle cleat bolted to the top flange of rafters and to columns. Figure 30 is an illustration of the bolted fixing of steel angle purlins to cleats with a short length of cleat for fixing along the length of a purlin and a longer length of cleat to make connection and provide fixing at butt ends of purlin connections. Similar angle section sheeting rails are bolted to cleats welded or bolted to columns.

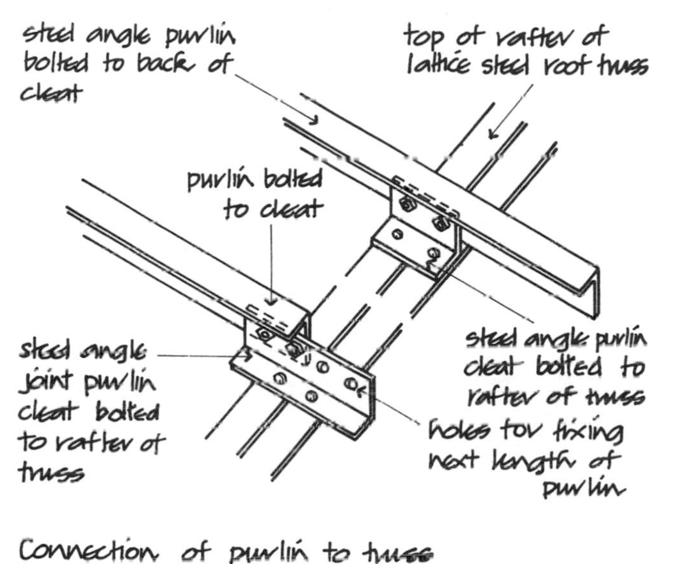

Fig. 30

Gable end wall sheeting is supported by and fixed to sheeting rails that are in turn fixed to steel gable posts of tee, channel or I-sections, bolted to a concrete pad, strip foundation, an upstand kerb or the concrete floor and fixed to the gable end truss as illustrated in Fig. 31. The gable end posts are fixed at centres to suit angle sheeting rails and gable wall sheeting.

Standard mild steel angles are not the most economical section for use as purlins and sheeting rails as the section is often considerably thicker than that required to support the dead weight and imposed loads on the roof and wall sheeting and the thickness of the standard angle is too great for the use of self-drilling fasteners that are used for fixing profiled steel sheeting.

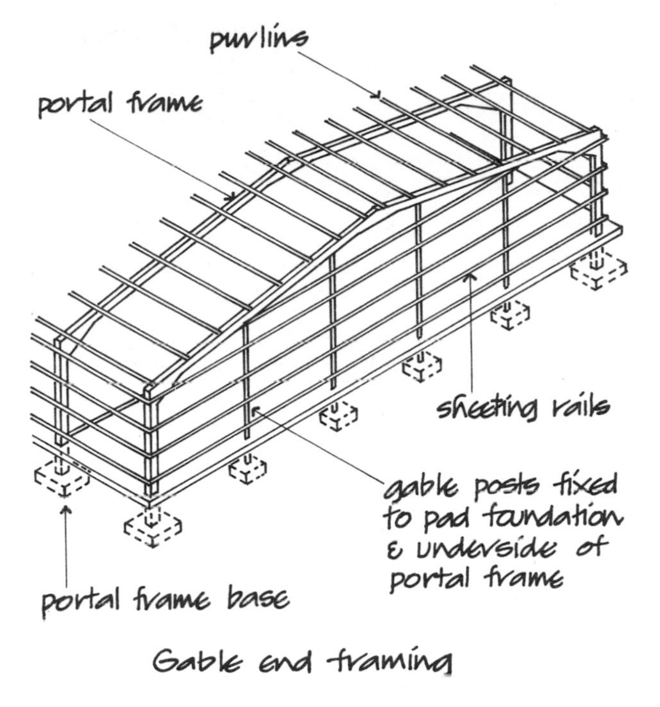

Fig. 31

Since about 1960 a range of galvanised, cold formed steel strip purlins and sheeting rails has been produced and designed specifically for the purpose. A range of standard sections and specifically designed purlins and rails is available.

The advantage of cold formed steel purlins and rails is economy in the use of material and flexibility in the design of the section to meet specific conditions of loading and use. The disadvantage of these purlins and rails is that due to the comparatively thin section of the material from which they are formed, various systems of anti-sag bars, braces and braces between purlins and rails are necessary to prevent gross distortion of the sections while sheeting is being fixed, distortion due to wind uplift and distortion due to the weight of side wall sheeting.

The sections most used are Zed and Sigma as illustrated in Fig. 32. More complex sections with stiffening ribs are also produced.

An advantage of the thin section of these purlins and rails is that it facilitates direct fixing of sheeting by self-tapping screws. The section of purlin and rail used depends on the type of profiled sheeting used. This determines the maximum spacing and the span between support from structural frames that subsequently determines the depth and section of the purlin and rail. Purlins and rails may be used in single or

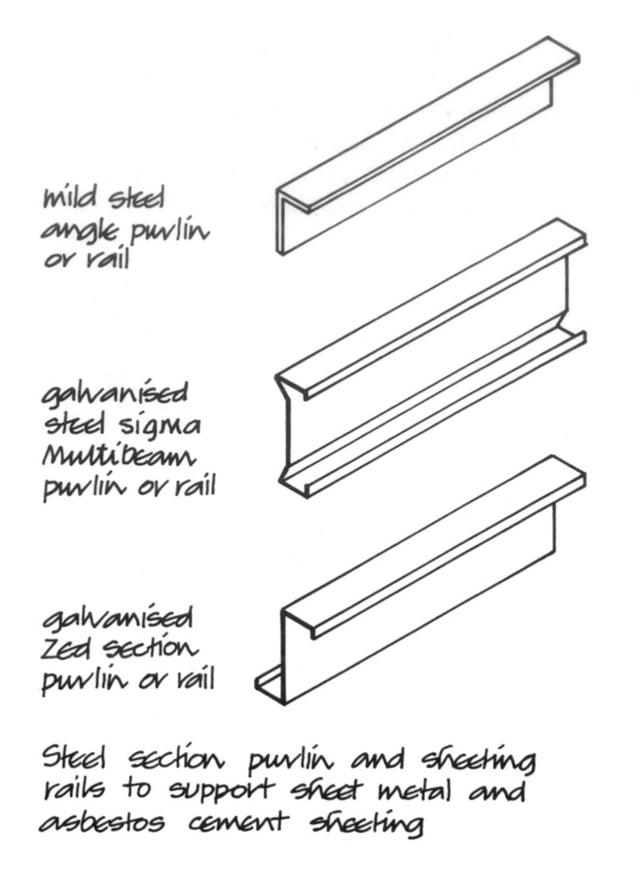

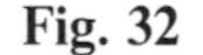
Fig. 32

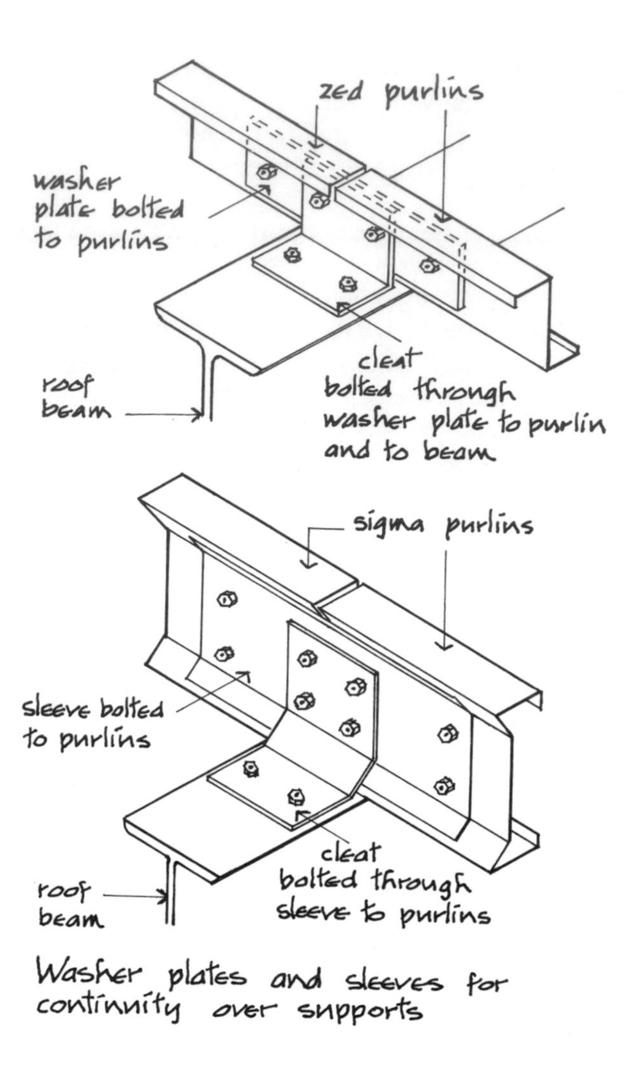

Fig. 33

double lengths between supports to which they are fixed with cleats to supports, washer plates and sleeves to provide continuity over supports. The typical cleats, washer plates and sleeves, illustrated in Fig. 33, are holed for bolts for fixing to purlins, rails and structural frame supports. Anti-sag bars are fixed between cold formed purlins to stop them twisting during the fixing of roof sheeting and to provide lateral restraint to the bottom flange against uplift due to wind pressure. When the sheeting has been fixed, the purlins derive a large measure of stiffness from the sheeting which acts as a roof membrane. Anti-sag bars should be used where the span of purlins, between support from the structural frames, exceeds 4.6 and at such intervals that the unsupported length of purlins does not exceed 3.8.

Anti-sag bars and apex ties are made from galvanised steel rod that is either hooked or bolted between purlins as illustrated in Fig. 34. The apex ties provide continuity over the ridge. For the system of anti-sag bars to be effective there must be some form of stiffening brace or strut at eaves as illustrated by the eaves brace in Fig. 34, which acts as a strut between the eaves purlin and eaves beam or structural framework.

The secret fixing for standing seam roof sheeting for low and very low pitched roofs does not provide lateral restraint for cold formed purlins either during sheet fixing or from wind uplift. With standing seam roof sheeting it is necessary to use a system of braces between purlins. These braces, which are manufactured from galvanised steel sections, are bolted between purlins as illustrated in Fig. 35, with purpose-made apex braces.

Sheeting rails are fixed across or between colums or the vertical members of frames at intervals to suit the profiled sheeting to be used. The Zed or Sigma section rails which are fixed with the flange of the section at right angles to the support are bolted to cleats and then bolted to the structural frame. A system of side rail

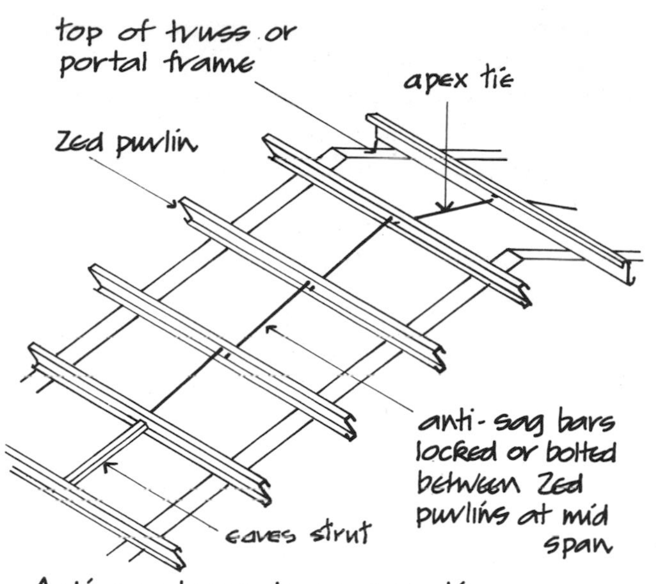

Fig. 34

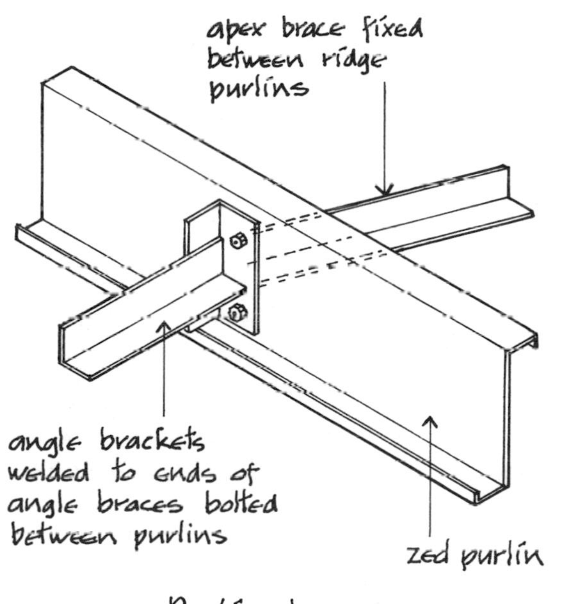

Fig. 35

struts is fixed between rails to provide strength and stability against the weight of the sheeting. For spans up to 6.0 one set of supports is used and above 6.0 two sets are used.

These side rail struts are fabricated from lengths of mild steel angle, to each end of which is welded a fixing plate which is bolted to the sheeting rails. In addition, a system of tie wires is fixed between the bottom two rows of rails and bolted to brackets fixed under cleats and supports. The fabricated struts, tie wires and clips are galvanised after manufacture. Fig. 36 is an illustration of side rails, struts and tie wires.

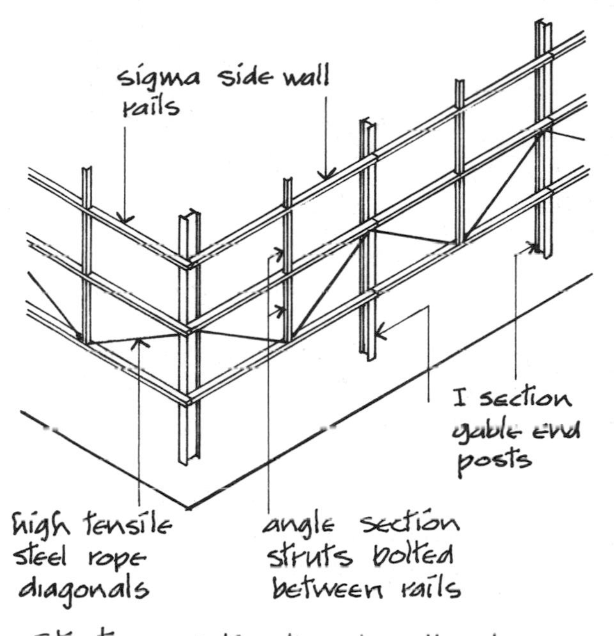

Fig. 36

As an alternative to angle iron or cold formed steel purlins, timber has been used for short- and medium-span purlins between structural roof frames. The durable, non-corrosive nature of timber allied to simplicity of cutting and fixing makes timber a practical and economic alternative to steel.

For economy in the use of materials, widely spaced roof frames are commonly used as support for deep profile and standing seam roof sheeting, laid over low and very low pitch roofs and fixed to either standard I-section or lattice steel purlins. The composite construction structural frames illustrated later in Fig. 53 employ I-section solid web, steel beam purlins which are sufficiently robust to need no lateral restraint.

Pre-cast reinforced concrete portal frames

For several years following the end of the Second World War (1945) there was a considerable shortage of structural steel in this country and it was then that the reinforced concrete portal frame came into common use for agricultural, storage, factory and other single-storey buildings.

A limited range of standard, pre-cast reinforced concrete portal frames was supplied for the economic benefit of repetitive casting in standard moulds and close control of mixing, placing and compaction of concrete that is possible in factory conditions. The immediate advantage of these building frames was that there was a ready supply of a limited range of standard frames that could rapidly be transported, erected and finished at an economic cost so avoiding the delays consequent on obtaining the necessary licence for the use of steel that was required at the time. The advantages of speed of erection and economy in the use of a limited range of standard sizes continued for some years after steel became more freely available.

The advantages of reinforced concrete portal frames are that they require no maintenance during the useful life of the building and the frame has a somewhat better resistance to collapse during fires than an unprotected steel frame. The principal disadvantage of these frames is that as they have to be formed in standard size moulds, for the sake of economy, there is only a limited range of sizes. The comparatively small spans that are practicable and the bulky somewhat unattractive appearance of the members of the frame have led to the loss of favour of this building system which is much less in use than it once was.

Due to the non-ductile nature of the principal material of these frames, i.e. concrete, the advantage of economy of section area gained by the use of the plastic method of design in the design of steel frames is considerably less with reinforced concrete. Because of the necessary section area of concrete and the cover of concrete to the steel reinforcement to inhibit rust and give protection to the steel reinforcement against damage during fires, the sections of the frames are large compared to steel frames of similar span. Damage to the frames and shrinkage cracks may rapidly cause rusting of the reinforcement particularly in wet and humid conditions. For convenience in casting, transport and erection on site, pre-cast concrete portal frames are generally cast in two or more sections which are bolted together on site either at the point of contraflexure in rafters or at the junction of post and rafter, or both, as illustrated in Fig. 38.

The point of contraflexure is that position along the rafters where negative or upward bending changes to positive or downward bending. At this point the member is presumed to be suffering no bending stresses so that structurally this is the soundest point to make a connection. Concrete portal frames are usually spaced at from 4.5 to 6.0 apart to support pre-cast reinforced concrete purlins and sheeting rails, cast in lengths to span between frames and hooked or bolted to the rafters and posts. As an alternative cold-formed steel Zed purlins and sheeting rails may be used for the fixing of profiled steel sheeting.

The bases of concrete portal frames are placed in mortices cast in concrete pad or strip foundations and grouted in position.

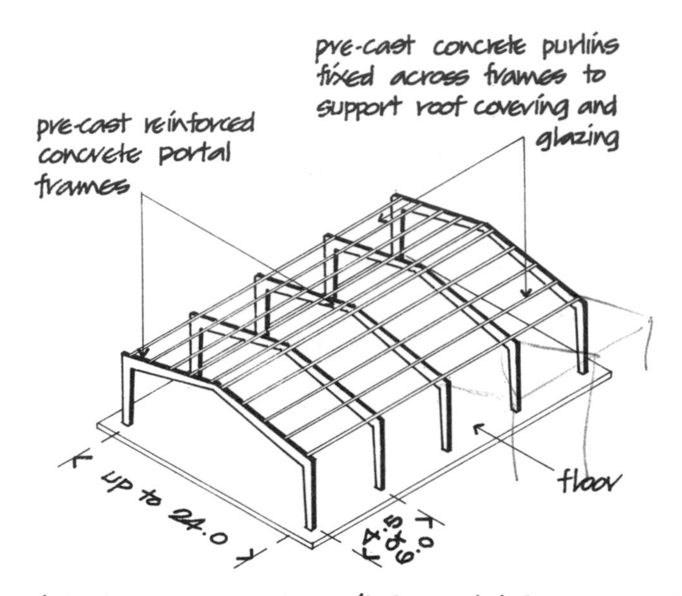

Fig. 37

Symmetrical pitch reinforced concrete portal frame construction

This is the most structurally efficient and most commonly used type of concrete portal frame.

It has been used for factories, warehouses, barns, sheds and single-storey places of assembly. Figure 37 is an illustration of a single bay symmetrical pitch pre-cast reinforced concrete portal frame. The slope of the rafters and spacing of purlins and sheeting rails is usually arranged to suit fibre cement or profiled steel sheeting.

Figure 38 is an illustration of the details of a two-bay symmetrical pitch concrete portal frame. It will be seen that the rafter, which is cast as one unit, is bolted to the

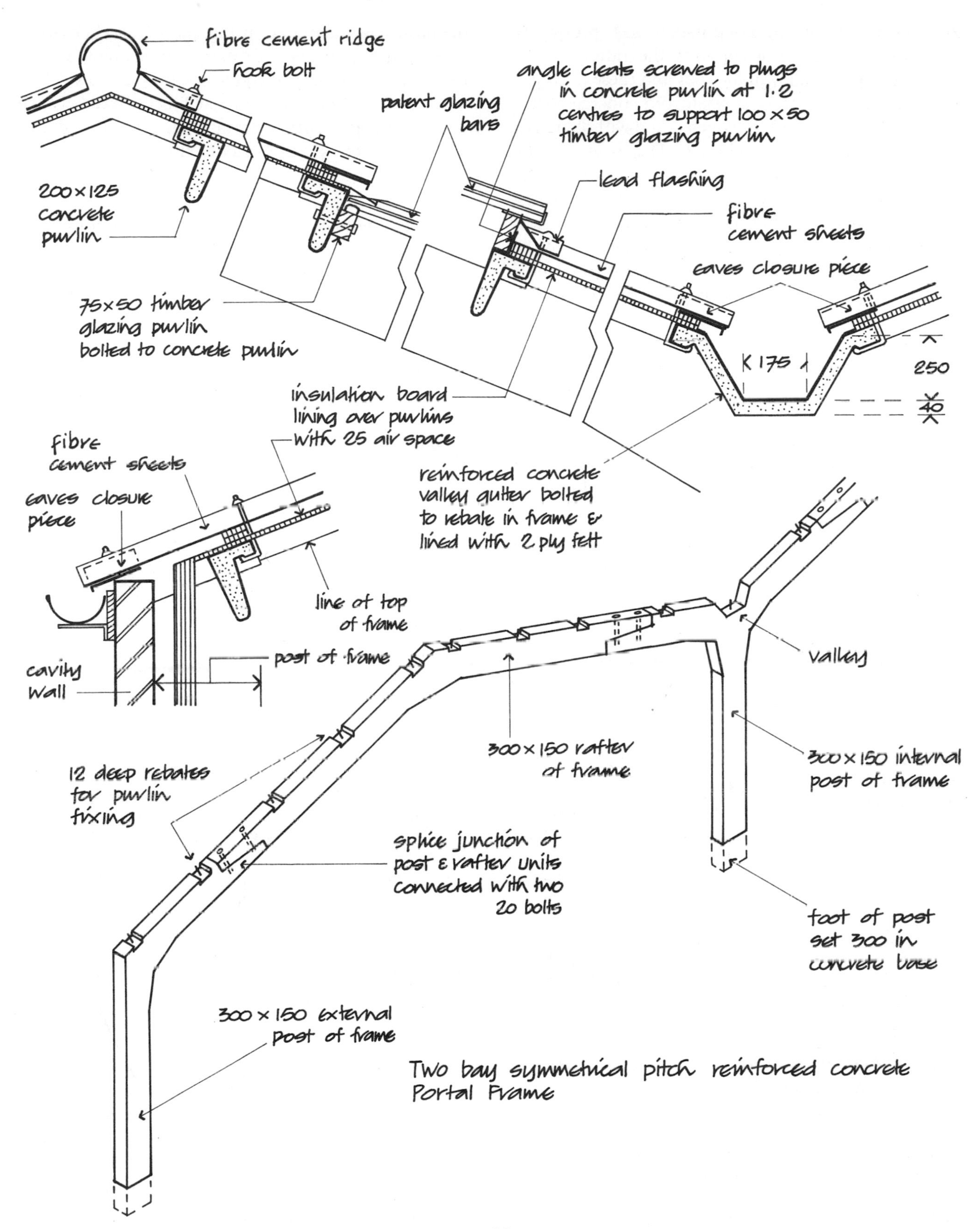
fibre cement ridge
hook bolt
angle cleats screwed to plugs in concrete purlin at 1.2 centres to support 100 × 50 timber glazing purlin
patent glazing bars
lead flashing
fibre cement sheets
eaves closure piece
200 × 125 concrete purlin
75 × 50 timber glazing purlin bolted to concrete purlin
175
250
40
insulation board lining over purlins with 25 air space
reinforced concrete valley gutter bolted to rebate in frame & lined with 2 ply felt
fibre cement sheets
eaves closure piece
line of top of frame
post of frame
cavity wall
valley
300 × 150 rafter of frame
300 × 150 internal post of frame
12 deep rebates for purlin fixing
splice junction of post & rafter units connected with two 20 bolts
foot of post set 300 in concrete base
300 × 150 external post of frame
Two bay symmetrical pitch reinforced concrete Portal Frame

Fig. 38

posts at the point of contraflexure as previously described. A single post supports the rafters of the frames below the valley in the roof and these posts are shaped to receive a pre-cast reinforced concrete valley gutter, bolted to the rafters, which is laid without fall to rainwater pipes and lined with felt. The spacing of the internal columns below valleys may be increased by the use of a pre-cast concrete valley beam to support every other internal roof frame. The bulky valley beam will obstruct clear head room and add considerably to the cost of the structure. The disadvantage of this multi-bay form of concrete portal frame is the number of comparatively bulky internal columns obstructing a free working area.

The pre-cast reinforced concrete purlins are usually of angle section with stiffening ribs and cast in lengths to span between portal frames. The purlins are fixed by loops protruding from their ends which fit over and are bolted to studs cast in the rafters, with the joint being completed with in-situ-cast cement- and sand-mortar, as illustrated in Fig. 39.

Corrugated fibre cement sheeting is hook bolted to the concrete purlins over an insulating lining laid over

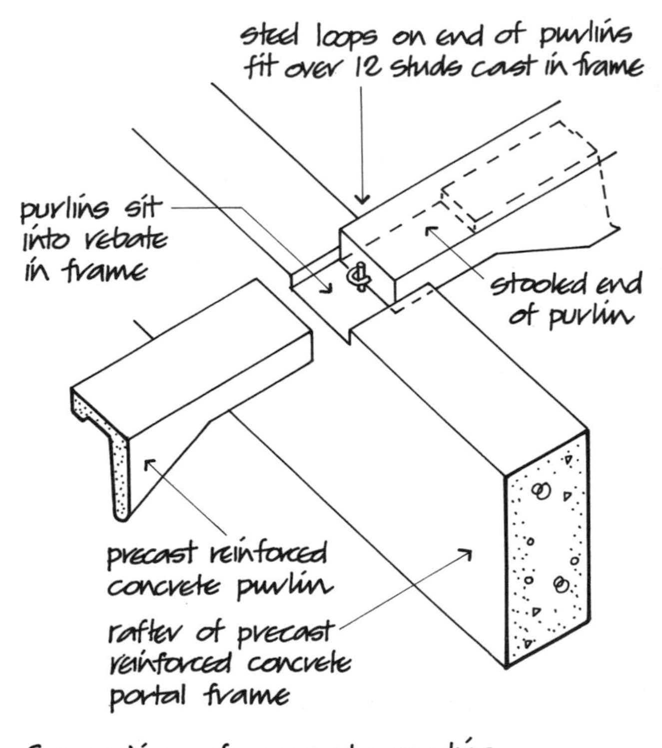

Connection of concrete purlins to concrete portal frame

Fig. 39

the purlins as illustrated in Fig. 38. As an alternative, profiled steel sheeting with an insulating lining may be fixed to Zed purlins bolted to the portal frames. Walls may be of solid brick or concrete blocks fixed between or across the posts of the portal frames, or fibre cement or profiled steel sheeting may be used.

North light pre-cast reinforced concrete portal frame construction

The most economical span for this profile of frame is up to about 9.0 to minimise the volume of roof space inside the frames and to avoid the large sections of frame that would be necessary with greater spans.

The south-facing slope is pitched at 22° and the north-facing slope at 60° to the horizontal. Figure 40 is an illustration of the frames of a typical two-bay north light concrete portal frame.

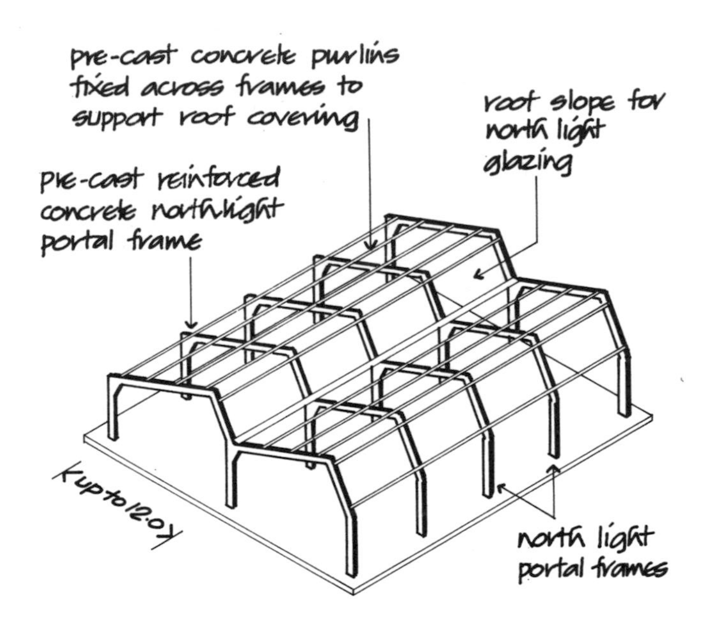

Two bay north light portal frames

Fig. 40

From Fig. 41 it will be seen that for convenience in casting and transport the rafter is cast in two sections which are bolted together at the point of contraflexure and in turn bolted to the posts. A pre-cast reinforced concrete valley gutter may be bolted to the frames as previously described.

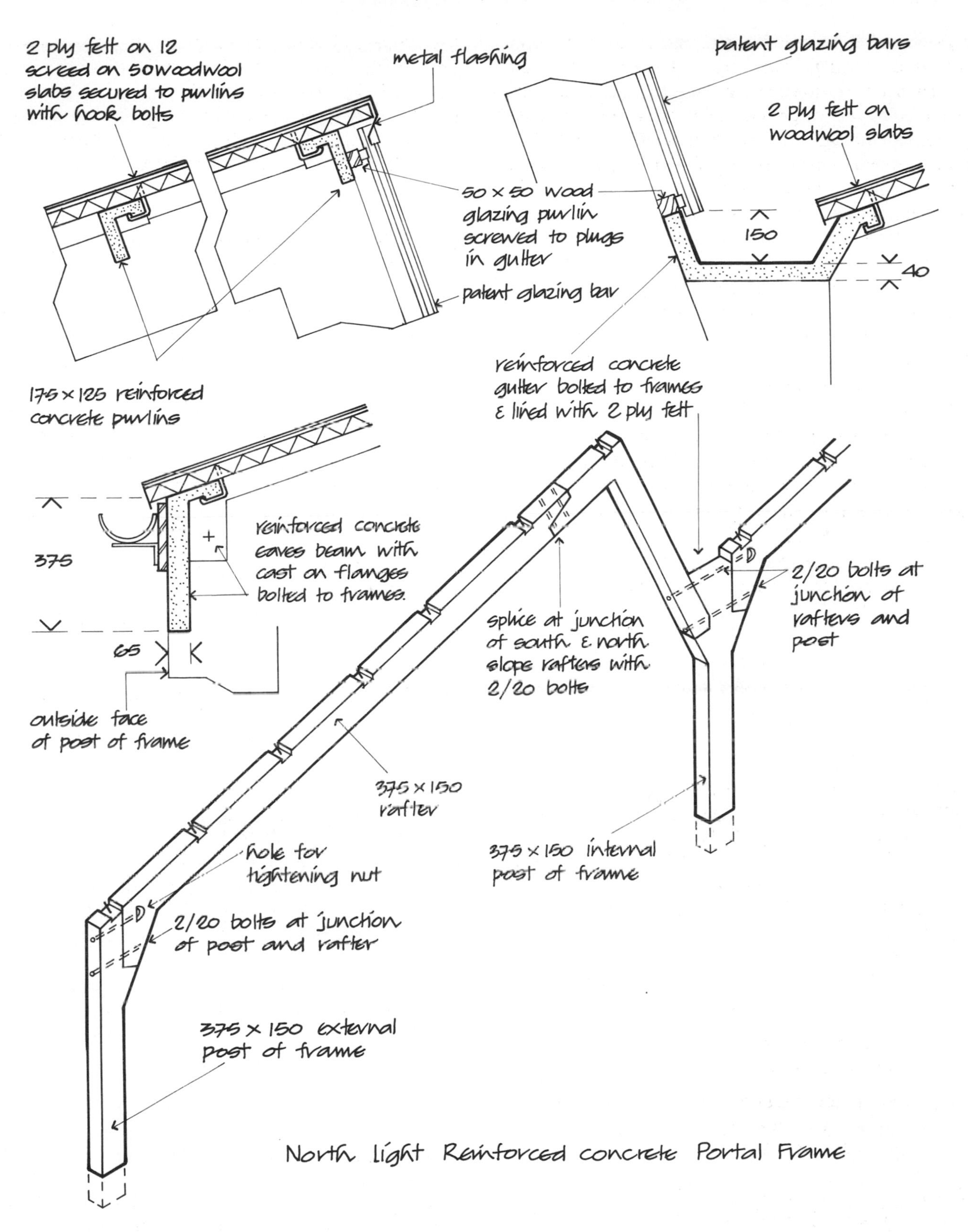
2 ply felt on 12 screed on 50 woodwool slabs secured to purlins with hook bolts
metal flashing
patent glazing bars
2 ply felt on woodwool slabs
50 x 50 wood glazing purlin screwed to plugs in gutter
150
40
patent glazing bar
175 x 125 reinforced concrete purlins
reinforced concrete gutter bolted to frames & lined with 2 ply felt
375
reinforced concrete eaves beam with cast on flanges bolted to frames.
65
outside face of post of frame
splice at junction of south & north slope rafters with 2/20 bolts
2/20 bolts at junction of rafters and post
375 x 150 rafter
hole for tightening nut
2/20 bolts at junction of post and rafter
375 x 150 internal post of frame
375 x 150 external post of frame
North light Reinforced concrete Portal Frame

Fig. 41

Pre-cast reinforced concrete purlins or steel Zed purlins are bolted to the rafters to support wood wool slabs, fibre cement or profiled steel sheets and north light glazing is fixed to timber purlins. In the example illustrated in Fig. 41 a pre-cast reinforced concrete eaves beam serves as purlin and provides support for the eaves gutter.

Because of the limited spans and the obstruction of many internal columns, this type of frame is much less used than it was.

Timber portal frames

Timber Up to about 1970 the conventional method of assessing the strength of timber, to be used structurally, was by visual examination of the surfaces of known specie of wood to assess a strength grade. An experienced grader could give a reasonable strength or stress grade. These visual grades tended to be on the conservative side, as the visual examination took no account of the density of the wood which has a large influence on both stiffness and strength. Once the relationship between stiffness and strength of timber had been established, it became practical to use non-destructive machines to measure the stiffness of timber as it passed through a machine to measure either the force required to produce a fixed deflection or to measure deflection caused by a known force at a particular point.

The majority of timber used for structural work is now machine-graded within nine strength classes and a wide range of stress grade/specie combinations from which suitable timber may be selected with confidence for structural use.

Fungal attack Timber has a natural resistance to fungal decay, which varies with the specie of timber, and is affected by the moisture content. Providing the moisture content of a timber is maintained at 20%–22% or less its natural resistance will protect it from fungal decay.

Insect attack The likelihood of attack by the most common form of beetle to attack wood in this country, the furniture beetle, is unpredictable. The attack, which is generally on internal dried sapwood, takes the form of holes bored along the long grain of the wood which may in time affect the strength of the wood. Insect attack, which is not as widespread as is generally believed, can be prevented by impregnating timber with an insecticide preservative. The house longhorn beetle may attack softwood in roof voids, where there is sufficient warmth, in an area of the Home Counties around London. In this area it is a requirement of the Building Regulations that softwood timber in roof voids be adequately protected.

Fire resistance Timber, which is a combustible material, is not easy to ignite in the sizes usual to buildings. Once ignited, timber burns very slowly and forms a protective layer of charcoal on its surface which insulates the remainder or residual section from the worst effects of fire.

It is possible to make a reasonably accurate estimate of the extent of the depletion or loss of timber in a fire and calculate the strength of the residual timber in supporting anticipated loads.

Surface spread of flame To limit rapid spread of fires in buildings it is a requirement of the Building Regulations that the surfaces of exposed elements have limited rates of surface spread of flame. The majority of softwoods used in buildings have a medium flame spread classification. In situations where there is a requirement for a low or very low rate of flame spread, the surface of softwood timber can be treated with flame retardants to achieve the necessary rate of flame spread.

Portal frames

Combinations of slender timber sections glued, or glued and nailed together, are used in portal frames for medium- and long-span roofs for such buildings as churches, assembly halls, sports halls and other single-storey structures where the timber portal frames are exposed for appearance sake. The advantages of timber as a structural material in this form are its low self-weight and the comparatively little maintenance required to preserve and maintain its strength and appearance, particularly where there are levels of high humidity as in swimming pools.

Symmetrical-pitch glued laminated timber portal These portal frames are usually fabricated in two sections for ease of transport and are bolted together at the ridge as illustrated in Fig. 42. These comparatively expensive portal frames are spaced fairly widely apart to support timber or steel purlins which can be covered with any of the sheet materials, slates or tiles.

The laminations of timber from which the portal is

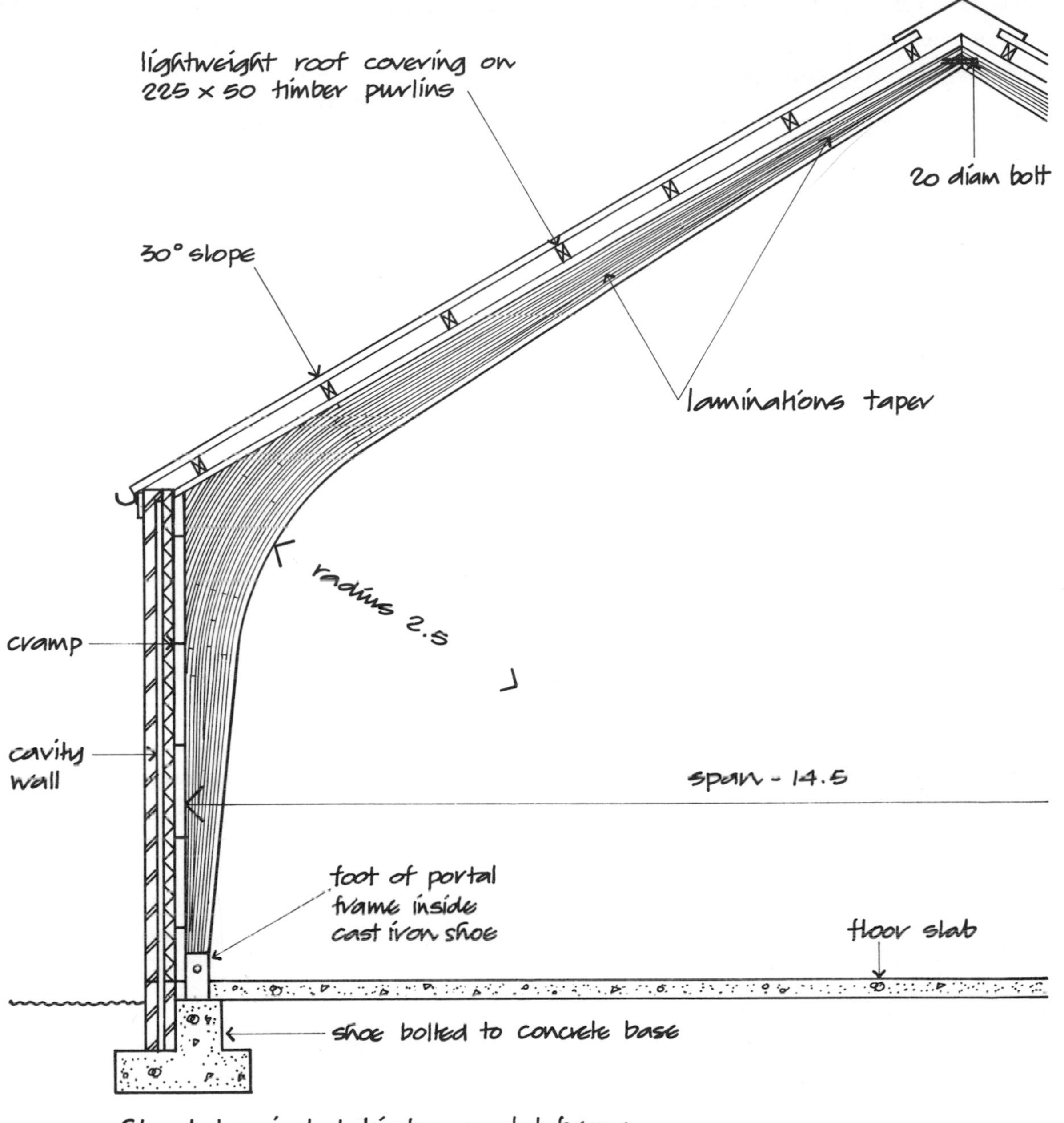

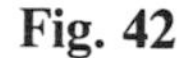
Fig. 42

made are arranged to taper in both the rafter and the post so that the depth is greatest at the knee, where the frame tends to spread under load, and slender at the apex or ridge and the foot of the post, where least section is required for strength and rigidity.

Because of their graceful arch-like appearance, glued laminated portal frames are used as much for appearance as practicality.

Flat glued-and-nailed timber portal The timber portal illustrated in Fig. 43 is a one-off design for an aircraft hanger. The flat portal frame is designed for the most economic use of timber and consists of a web of small section timbers glued together with the top and bottom booms of glued laminate with webb stiffeners. The portal frames are widely spaced to support metal decking on the roof and profiled sheeting on the walls.

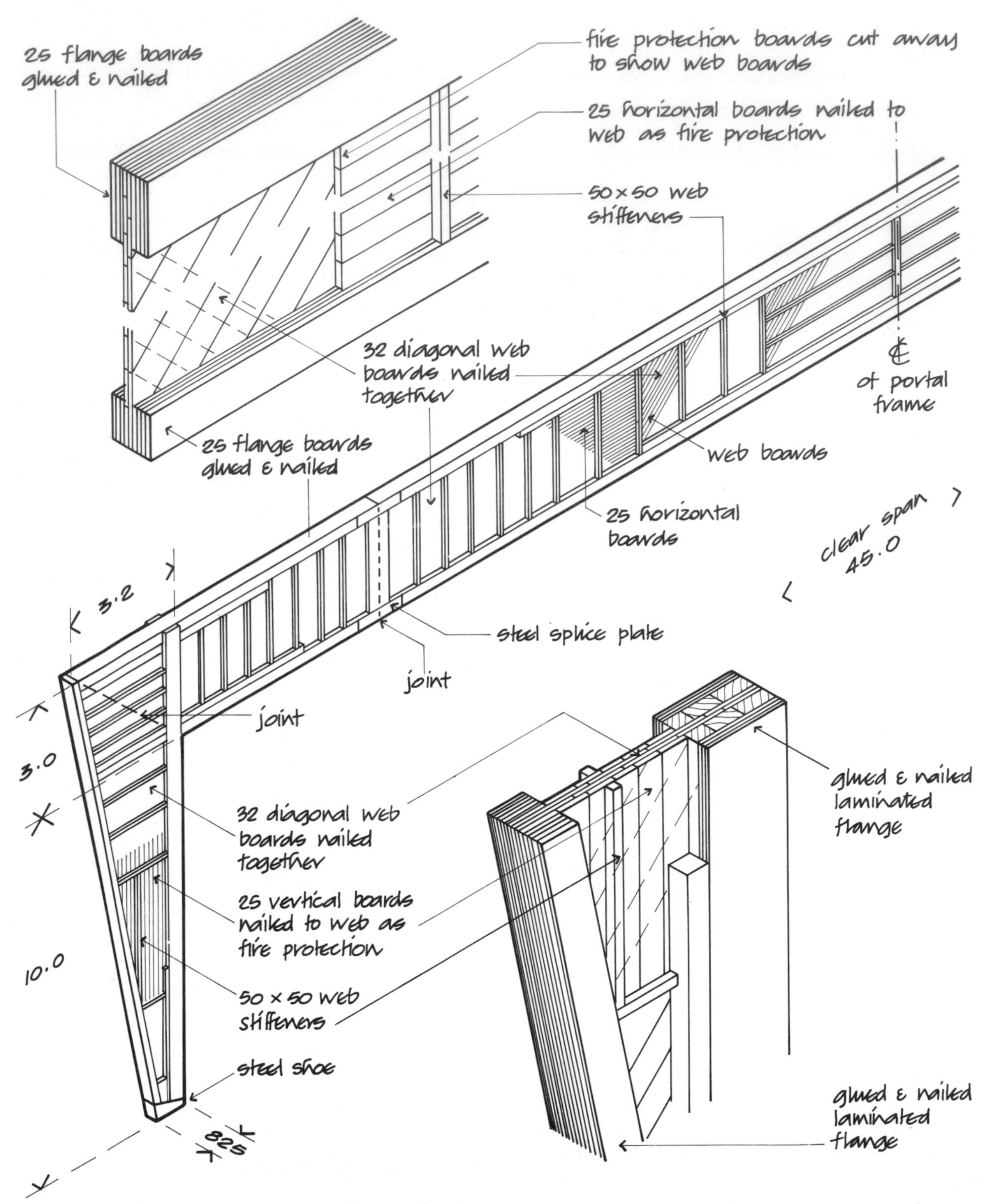
25 flange boards glued & nailed
fire protection boards cut away to show web boards
25 horizontal boards nailed to web as fire protection
50 x 50 web stiffeners
32 diagonal web boards nailed together
℄ of portal frame
25 flange boards glued & nailed
web boards
25 horizontal boards
clear span 45.0
3.2
steel splice plate
joint
joint
3.0
32 diagonal web boards nailed together
glued & nailed laminated flange
25 vertical boards nailed to web as fire protection
10.0
50 x 50 web stiffeners
steel shoe
825
glued & nailed laminated flange
Glued and nailed timber Portal Frame

Fig. 43

This structural form was chosen both for appearance and the long span structure which is lightweight, free from maintenance and has adequate fire resistance.

FLAT ROOF FRAME CONSTRUCTION

The design of buildings is often more subject to the dictates of fashion than economy in first cost, utility and maintenance. The appearance of single-storey buildings, such as factories covering large floor areas, is influenced by the profile of the roof structure. For some years it was fashionable to adopt the strong horizontal roof lines of a flat roof structure on the grounds that it was modern, rather than the more economical single or multi-bay pitched roof profile. Over the years these large areas of flat roof have come into disrepute due to the failure of flat roof coverings which have not remained watertight because of movements of the covering relative to the decking and structure and failures at junctions to parapets and rooflights.

Recent improvements in flat roof coverings to enhance strength and elasticity of the material and delay brittle hardening by oxidation, together with improvements in design detailing to allow for movements of the weather surface and roof support, have improved the useful life of flat roofs to compare favourably with profiled sheet coverings.

Medium- and long-span flat roof structures are less efficient structurally and therefore somewhat more expensive than truss, lattice or portal frames. The main reason for this is the need to prevent too large a deflection of the flat roof structure under load to the accepted 1/250 of span and to limit deflection to prevent ponding of rainwater on flat roofs. For these reasons flat roof beams and girders have to be deeper than is necessary for strength alone. Ponding is the word used to describe the effect of rainwater lying in the centre of flat roofs, where deflection under load is greatest, in the form of a shallow pool of water that cannot drain to the rainwater outlets. A static pool or pond of water will plainly penetrate faults in the roof covering more readily than water running off to gutters and outlets. To avoid ponding the roof surface has to have a positive fall to outlets under load. A fall or slope of at least $2\frac{1}{2}°$ is considered as an absolute minimum, so that at mid-span there is some fall to boundary outlets.

An advantage of flat roof frames is that there is comparatively little unused roof space to be heated.

Main and secondary beam flat roof construction

Figure 44 is an illustration of a single-bay flat roof structure with solid web I-section main beams supported by steel columns with I-section secondary beams between the main beams. To provide a positive fall (slope) to eaves gutters at each side of the roof, it is necessary to fix tapered depth bearers of steel or timber across the secondary beams to provide the necessary fall from the centre to each side of the roof for the decking and weathering finish. This heavy construction is not structurally efficient because of the considerable depth required in the main beams to limit deflection under load. This type of structure is used for single-bay short- or medium-span roofs where the main beams are used to provide support for travelling cranes and other lifting gear. Because services have to be incorporated below the solid web beams of the roof structure there is a considerable increase in the volume of unused roof space that may have to be heated.

Lattice beam (girder) flat roof construction

The required depth of beams for flat roof construction is determined by the need to limit deflection under load

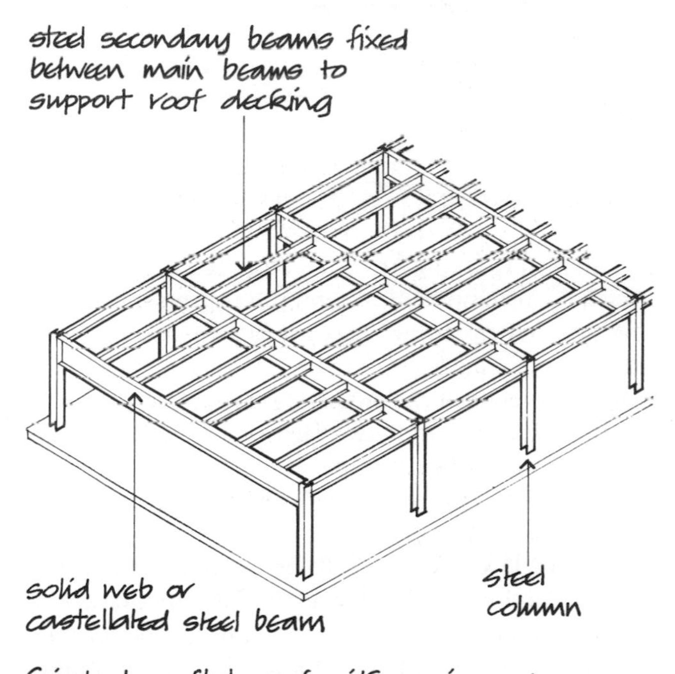

Fig. 44

and by the rigidity of depth more than the weight of the material in the beam. A lattice beam or girder of comparatively small section members, fixed between the top and bottom boom, at once provides adequate stiffness and economy in the use of materials and low self-weight. Because lattice beams and girders have to be fabricated it is generally an advantage to taper the top boom to provide a positive fall or slope for the roof decking and weathering. A taper or low pitch lattice beam or girder may be specially fabricated or one of the standard beams, with either one way or two way taper, can be used.

The terms beam and girder are used in a general sense to describe a lattice construction, 'beam' being used for comparatively small depths such as those used for roofs, and 'girder' for those of appreciable depth used to support heavier loads such as those in bridge construction.

Short-span beams that support comparatively light loads may be constructed from cold-formed steel strip top and bottom booms with a lattice of steel rods welded between them as illustrated in Fig. 45. The top and bottom booms are formed as 'top hat' sections designed to take timber inserts for fixing roof decking and ceiling finishes. These beams are finished with a stove-enamelled primer ready for painting. These standard beams are considerably cheaper than one-off beams, through the economy of mass production.

The majority of lattice beams used for flat and low pitch roofs are fabricated from hollow round and rectangular steel sections. A lattice of hollow round sections is welded to hollow rectangular section top and bottom booms, with end plates for fixing to supports. Hollow rectangular section booms are preferred for the economy in making straight, oblique cuts to the ends of the lattice round sections and the convenience of roof fixings. Where round section booms are used it is necessary to make a more complicated oblique mitre cut to the ends of the lattice members.

Both for flat and low pitch roofs it is generally convenient to fabricate taper lattice beams with the top booms with a one-way or two-way slope to provide the necessary falls or slope to drain to rainwater outlets. For most low pitch roofs to be covered with profiled sheeting a slope of 6° is provided. The lattice beams are either hot dip galvanised, stove enamel primed or spray primed after manufacture. Figure 46 is an illustration of a typical lattice roof beam fabricated from hollow steel sections.

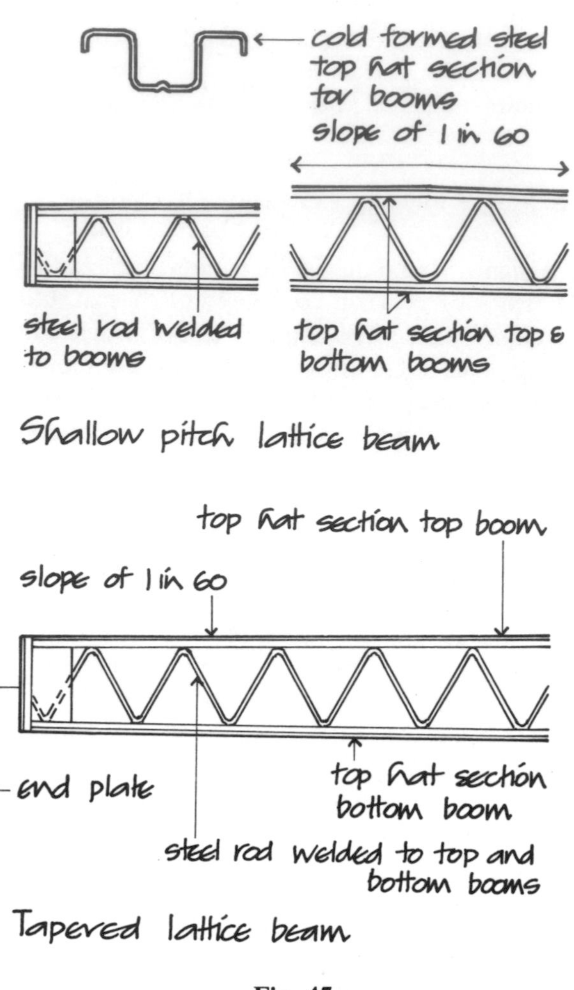

Fig. 45

V-beam flat roof construction

A system of V-section standard grid lattice beam is supported by end lattice beams supported on steel columns. The V-section or prismatic beams are fabricated from tubular steel sections welded together. The V-beams are spaced to support metal decking across the whole of the roof or the V-beams can be spaced apart to suit continuous or separate rooflights. Figure 47 is an illustration of a single-bay, single-storey lattice V-beam structure. With standard section, standard span lattice V-beams, a reasonably economic single- or multi-span flat roof structure can be built.

Space grid flat roof construction

A two-layer space deck constructed of a grid of standard units is one of the commonly used flat roof

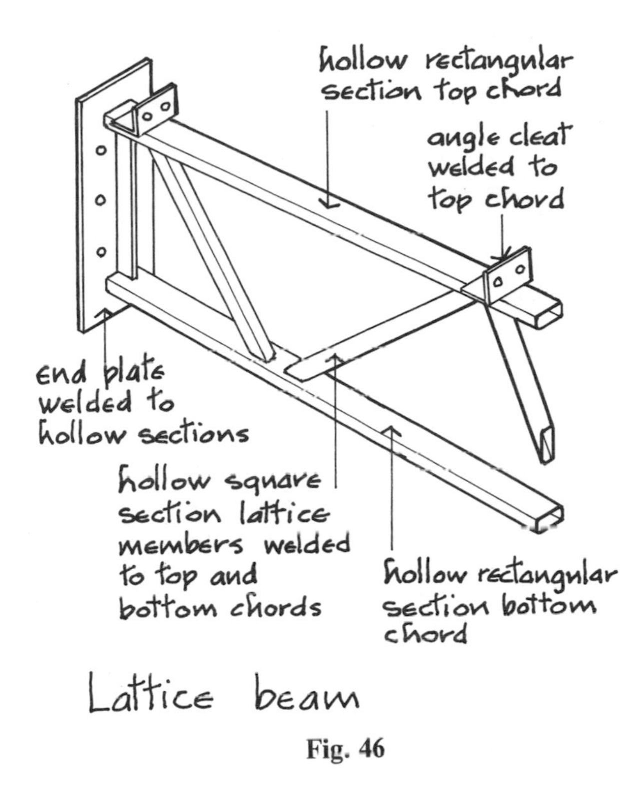

Fig. 46

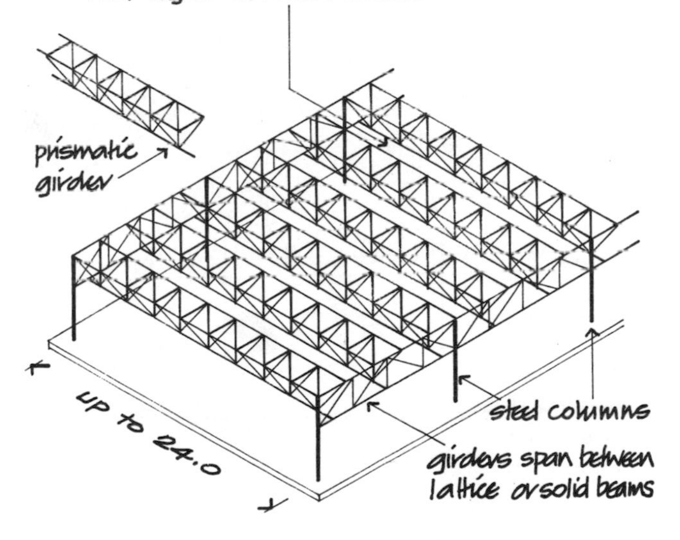

Fig. 47

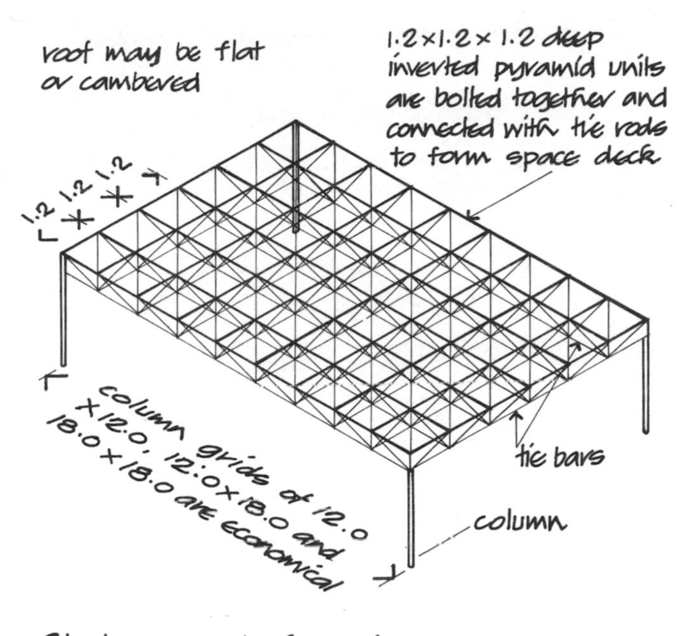

Fig. 48

structures for single-storey buildings such as sports halls, shopping centres, leisure halls, factories and other buildings where it is convenient to have the whole floor area free of obstructing columns. Figure 48 is an illustration of a single-storey structure with a space deck flat roof supported on steel columns.

The space deck is assembled on site from standard space deck units, each in the form of an inverted pyramid with a steel angle tray base, tubular diagonals welded to the tray and a coupling boss as illustrated in Fig. 49. The space deck units are bolted together through the angle trays and connected with tie bars through the coupling bosses. The tie bars which have right- and left-hand threads can be adjusted to give an upward camber to the top of the deck to allow for deflection under load and to provide a positive fall to the roof to encourage the run-off of rainwater and so avoid ponding.

Space deck roofs may be designed as either two-way spanning structures with a square column grid or as one-way spanning structures with a rectangular column grid. Economic column grids are 12.0 × 12.0, 18.0 × 18.0 and 12.0 × 18.0. Various arrangements of the column grid are feasible with also a variety of roof levels, canopies and overhangs.

The advantages of the space deck roof are the comparatively wide spacing of the supporting columns, economy of structure in the use of standard units and

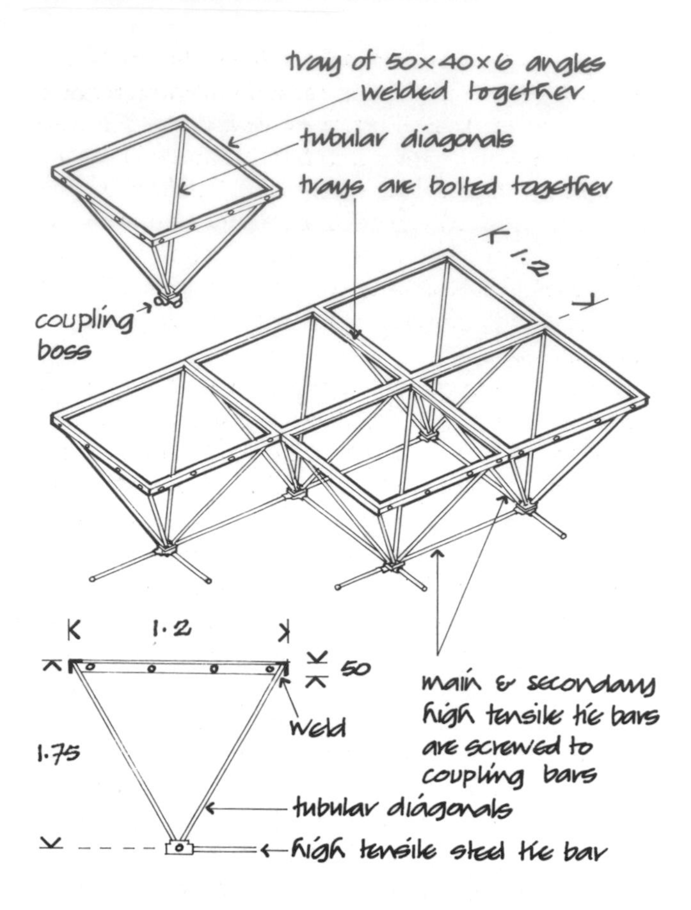

Fig. 49

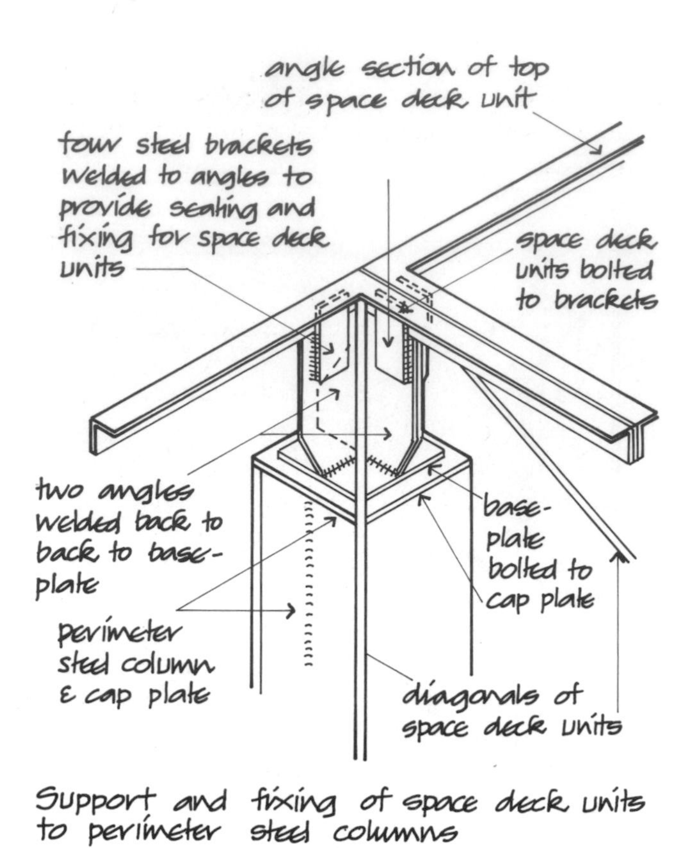

Fig. 50

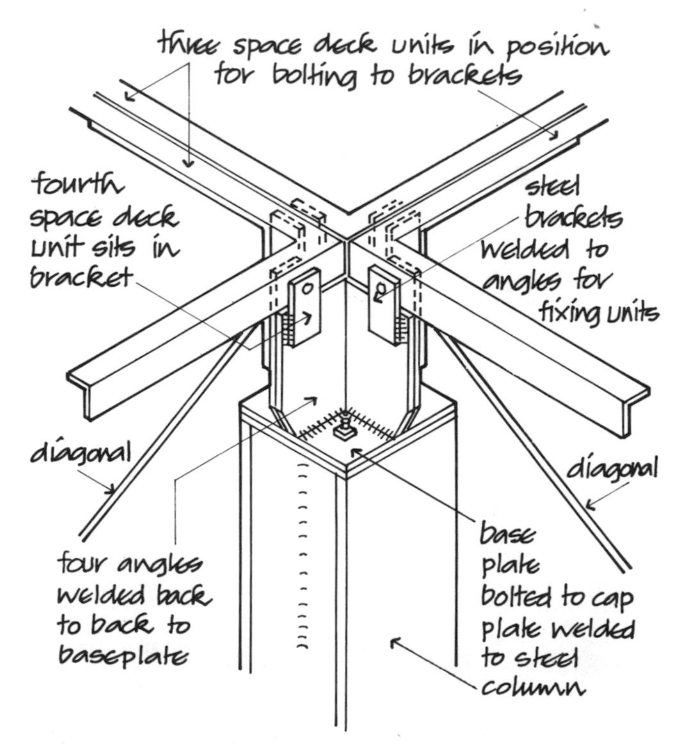

Fig. 51

speed of erection. The disadvantage of the space deck is the great number of lattice members that will collect dust and require careful maintenance to inhibit rust.

The roof of the structural space deck may be covered with steel decking, insulation and one of the flat roof weatherings. Rooflights can be accommodated within one or more of the standard space deck units.

Steel columns supporting the space deck are usually connected to the units at the junction of the trays of the units. Figure 50 is an illustration of the junction of a column at the perimeter of the structure. A steel cap plate is welded to the cap of the column to which a seating is bolted. This seating of steel angles has brackets welded to it into which the flanges of the trays fit and to which the trays are bolted. Likewise a seating is bolted to a cap plate of internal columns with brackets into which the flanges of the angles of four trays fit, as shown in Fig. 51.

The space deck can be finished with either flat or sloping eaves at perimeter columns or the deck can be cantilevered beyond the columns as an overhang. Figure 52 illustrates a two-deck unit overhang and a fixed base with the columns bolted to a concrete pad foundation.

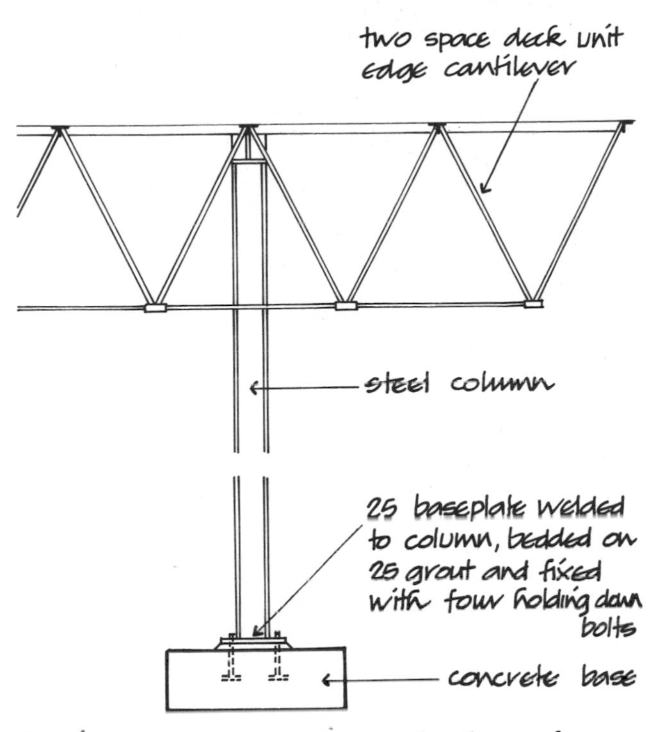

Fig. 52

COMPOSITE FRAME CONSTRUCTION

For many years in the second half of the twentieth century the majority of structural frames were constructed with reinforced concrete, in part due to an initial shortage of steel and in part due to the fashion for the newer material concrete. The somewhat better resistance to damage by fire of reinforced concrete, as compared to steel alone, also played some part in the preference for concrete.

Most reinforced concrete structural frames are constructed as concrete cast in-situ inside temporary falsework of timber or steel as a mould and as temporary support for the initially wet concrete. Complex and labour-intensive systems of falsework and support have to be erected and maintained in place until the concrete gains sufficient strength to be self-supporting and then dismantled (struck) and erected again and then struck at each floor level. This is a costly and somewhat illogical way of building.

With changes in Building Regulations which accepted lightweight dry linings and mineral fibre and cement coatings to steel as fire protection, and a change from requirements for protection of the building to requirements for safety in escape from buildings in case of fire, reinforced concrete no longer enjoyed an advantage over steel. In consequence, steel is now as much used for structural frameworks as reinforced concrete and is more generally used for single-storey buildings.

During recent years manufacturers equipped to precast a range of reinforced concrete structural beams and columns, under carefully controlled factory conditions, have successfully offered a comprehensive design, manufacture and erection service for both single- and multi-storey frames of pre-cast reinforced concrete frames and lattice steel roof beams, at prices competitive with structural steel frames. The advantages of these composite frames are that the reinforced concrete columns and beams will require little maintenance other than occasional washing for appearance and will provide better resistance to damage by fire than steel alone, where good sense or insurance requirements seek protection of valuable contents. Lattice steel beams are used for roofs, in this composite form of construction, where there is now no requirement for fire resistance.

The pre-cast reinforced concrete columns and beams are cast and compacted under closely controlled conditions in the factory. Frame joints and base fixings are cast in as necessary, and the exposed faces of the frame can be smooth and dense, so that no maintenance is required during the useful life of the building, or finished with a variety of finishes.

A variety of shapes for columns, beams and structural frames is practical and reasonably economic where repetitive casting of several like-members is called for.

Figure 53 is an illustration of a typical two-bay, single-storey composite frame structure. The pre-cast reinforced concrete columns, which have fixed bases, serve as vertical cantilevers to take the major part of the loads from wind pressure. Steel brackets, cast into the column head, support concrete and lattice steel roof beams. Concrete or lattic steel spine beams under the roof valley provide intermediate support for every other roof beam.

The top of the lattice steel roof beams, which are pitched at 6° to the horizontal, support low pitch, profiled steel roof sheeting. Fixing slots or brackets cast into the columns provide a fixing and support for sheeting rails for profiled steel cladding.

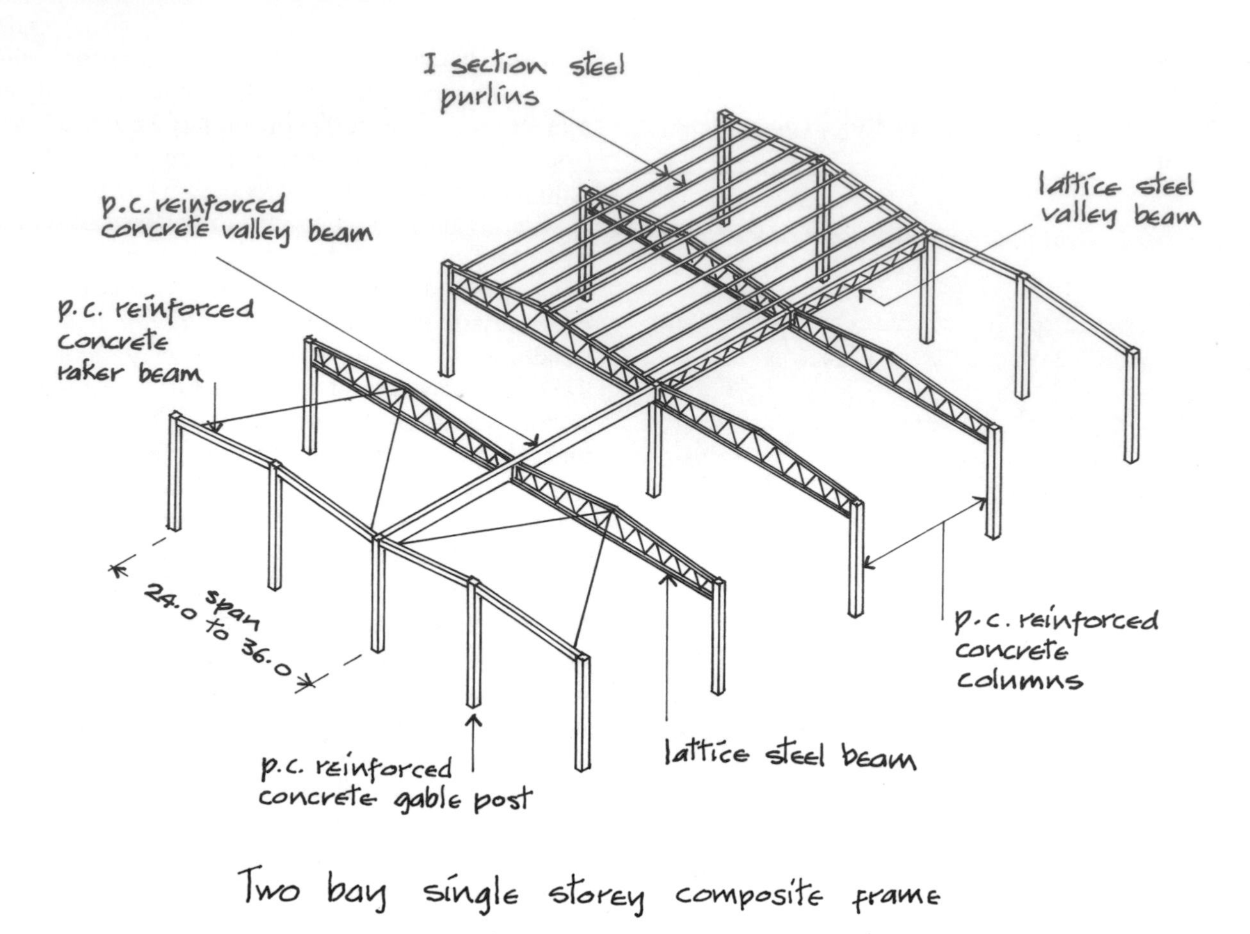
I section steel purlins
lattice steel valley beam
p.c. reinforced concrete valley beam
p.c. reinforced concrete raker beam
span 24.0 to 36.0
p.c. reinforced concrete columns
lattice steel beam
p.c. reinforced concrete gable post
Two bay single storey composite frame

Fig. 53

CHAPTER TWO

ROOF AND WALL CLADDING, DECKING AND FLAT ROOF WEATHERING

FUNCTIONAL REQUIREMENTS

The functional requirements of roofs and walls comprise:

Strength and stability
Resistance to weather
Durability and freedom from maintenance
Fire safety
Resistance to the passage of heat
Resistance to the passage of sound
Security.

Strength and stability

The strength of roof and wall cladding and roof decking depends on the properties of the materials used and their ability to support the self-weight of the cladding plus the anticipated wind and snow loads, between supporting purlins, rails, bearers and beams.

The stability of cladding and decking depends on the:

(1) Depth and spacing of the profiles of sheeting and decking
(2) Composition of the materials and thickness of the boards and slabs used for decking
(3) Ability of the materials to resist distortion due to the:
  (a) Wind pressure
  (b) Wind uplift
  (c) Snow loads
  (d) Weight of personnel engaged in fixing and maintaining the roofs.

The strength and stability of the comparatively thin sheets of steel or aluminium used for profiled sheeting derive principally from the depth and spacing of the profiles, from the shallow depth of corrugated sheet for small spans to the considerable depth of deep trapezoidal profiles and standing seams for medium to large spans between supports. Longitudinal and transverse ribs, to deep profile sheeting and decking, provide additional rigidity against buckling due to the distortion caused by point loads and the sometimes very considerable wind pressure and uplift.

Steel roof cladding sheets fixed across a structural frame act as a diaphragm which contributes to the stability of the frames in resisting the racking effect of the considerable lateral wind forces that act on the sides and roofs of large buildings. The extent of the contribution of the sheet covering to the stability of the frames depends on the thickness of the sheets, the strength of the fasteners used to fix the sheets and the strength of the sheets in resisting the tearing effect of the fasteners fixed through it. From a calculation of the likely maximum wind forces acting on a building, cladding sheets of adequate thickness to resist tearing away from fasteners, adequate profile to resist buckling and the required section and spacing of fasteners can be selected so that the sheeting will act in whole or part as a diaphragm to resist wind pressure on the building.

The comparatively thick corrugated and profiled fibre cement cladding sheets, used for short and medium-span support between purlins, rails and beams, have adequate strength in the depth of the profiles for the anticipated loads and rigidity in the thickness of the material to resist distortion and loss of stability over the moderate spans between supports.

Resistance to weather

The traditional roofing materials, tile and slate, effectively resist the penetration of rainwater to the interior by the run-off of rainwater down the slope over at least two thicknessses of tile or slate. The great advantage of these traditional materials is that the small units of slate or tile will accommodate the range of moisture and thermal and structural movements common to exposed roofs, without damage for the life of the majority of buildings and without suffering damage or deterioration, providing the slope of the roof is adequate and sound tiles and slates are securely fixed.

Large sheets of profiled steel and aluminium serve to resist the penetration of rainwater, through the impermeability of these metals to water, which runs down the slope of the roof over the necessary end and side laps of the sheets. The least slope of the roof is dictated by the end lap of the sheets necessary to prevent water finding its way up the slope and between closely fitting sheets. Thermal and structural movements across the roof are accommodated by the profiles that will allow for normal contraction and expansion and accommodated down the slope by some movement at fixings at end laps. Where long sheets are used, the secret fixings of standing seams will allow adequate movement.

Profiled metal sheets are usually fixed with screws driven through the troughs of the profiles to steel purlins and sheeting rails. Integral steel and neoprene washers on the screw head effectively seal the perforation of the sheet against penetration of rain. Recently it has become practice to fix profiled metal sheeting through the ridge of the troughs. This requires some care in driving the screw home to find a secure fixing without driving the screw so firmly that it distorts the profile. The advantage of this method of fixing is that the perforation of the sheet is less exposed to rainwater that will tend to run down the troughs. Standing seams to the long edges of sheets provide a deep upstand as protection against rain penetration at vulnerable long edges, particularly with very low pitch roofs. Secret fixings that do not require perforation of sheets accommodate thermal movements and are not visible for appearance sake.

Profiled cladding for walling is usually fixed through the troughs of the profile for ease of fixing and where the screw heads will be least visible.

The more solid, thicker profiled fibre cement sheets will resist the penetration of rainwater through the density of the sheet material, the slope of the roof and the end and side laps. These sheets, which will absorb some rainwater, should be laid at a pitch of not less than 10° to avoid the possibility of frost damage. The cladding will accommodate moisture, thermal and structural movement through the end and side laps and the comparatively large fixing holes for screws or hook bolts. Because of the thickness of these sheets they do not make a close fit at end laps through which there will be considerable penetration of wind under pressure. Flat roof weathering membranes which resist the penetration of rainwater through the impermeability of the two-, three- or single-ply membranes and the sealed joints or continuity of the membrane will, in time, harden and no longer retain sufficient elasticity or tensile strength to resist the very considerable thermal movements common to flat roof coverings laid over effective insulation materials.

Durability and freedom from maintenance

The durability of coated, profiled steel sheeting depends initially on some care in handling and fixing. The thin sheets may be distorted and damaged by careless handling and subject to damage through careless fixing that may lead to corrosion of exposed steel around fixing holes and distortion of sheets by fixings driven home too tightly. The edges of sheets should be painted or coated to provide protection of the steel exposed at edges of sheets.

For roofing and side wall cladding, the durability of profiled steel sheets depends on climate and the colour of the coating material. Sheeting on buildings close or near to the coast and buildings in heavily polluted industrial areas will deteriorate more rapidly than those inland. Sheeting with light coloured coatings is more durable than that with dark coloured coatings due to the effect of ultraviolet light on dark hues and the increased heat released from solar radiation on absorptive dark coatings.

The early deterioration of coated sheets is due to the formation of a white layer on the exposed surfaces of the organic coating that takes the form of an irregular, chalk-like coating that may become apparent, particularly with dark colours, some 10 to 25 years after fixing. This coating, which is termed chalking, does not adversely affect the protective properties of the coating. It may, however, so affect appearance as to be unacceptable, particularly on darker shades of colour. The coat of chalking can be removed by washing and lightly brushing the whole surface of the sheeting, which can then be overpainted with a primer and silicone, alkyd paint to improve appearance. After some years the coating may craze in the form of irregular, interconnected cracks and lose adhesion to the steel sheet. Dark coloured sheets are likely to suffer this type of defect sooner than light coloured sheets. A remedy for this defect is to strip affected areas of coating, clean and prime exposed steel surfaces and paint the whole of the cladding. At this stage of deterioration of the surface of the sheets it is probably more economical and certainly far more satisfactory to strip the whole of the sheeting and replace it with new. Because of the need for regular inspection and comparatively early remedial maintenance, organic coated steel sheeting is, at best, a short- to medium-term material with a useful life of 25 years in the most favourable conditions and as little as ten years before costly maintenance is necessary.

Asbestos cement and fibre cement sheeting does not corrode or deteriorate for many years providing it is laid at a sufficiently steep pitch to shed water. Because of the brittle nature of the material it is liable to damage by knocks or undue pressure from those walking carelessly over the surface. The coarse texture of the material readily collects dirt which is not washed away by rain and the irregular dirt staining of this sheet is not generally accepted as an attractive feature.

Flat roof weathering membranes, laid directly over insulation, suffer very considerable temperature variations between day and night due to the texture and colour of the materials that absorb solar radiation. The heat released is retained due to the insulation below. In consequence there is considerable expansion and contraction of the membrane between day and night which in time may cause the weathering layer or layers to tear. Solar radiation also causes oxidation and brittle hardening of bitumen saturated or coated materials which in time will no longer be impermeable to water.

Single skin membranes of synthetic rubber-like materials will, for some years, accommodate thermal movements by the inate elasticity of the material. In time, due to brittle hardening and gradual loss of elasticity, these materials will fail.

The durability of a weathering membrane in an inverted or upside down roof is much improved by the layer of insulation which is laid over the membrane protecting it from the destructive effects of solar radiation to a considerable extent.

The useful life of bitumen impregnated felt membranes is from 10 years, for organic fibre felts up to 20 years and for high performance felts up to 25 years. The useful life of the latter materials used in an inverted or upside down roof can reasonably be doubled.

Over the years, mastic asphalt as a weathering membrane will oxidise and suffer brittle hardening which, combined with thermal movements, will give this material a useful life of up to 20 years.

Fire safety

The requirements from Part B of Schedule 1 to the Building Regulations, 1991 are concerned to:

(a) Provide adequate means of escape
(b) Limit internal fire spread (linings)
(c) Limit internal fire spread (structure)
(d) Limit external fire spread
(e) Provide access and facilities for the Fire Services.

Fire safety regulations are concerned to ensure a reasonable standard of safety in case of fire. The application of the regulations, as set out in the practical guidance given in Approved Document B, is directed to the safe escape of people from buildings in case of fire rather than the protection of the building and its contents. Insurance companies that provide cover against the risks of damage to the buildings and contents by fire will generally require additional fire protection.

Means of escape The requirement from the Building Regulations is that the building shall be designed and constructed so that there are means of escape from the building in case of fire to a place of safety outside the building. The main danger to people in buildings, in the early stages of a fire, is the smoke and noxious gases produced which cause most of the casualties and may also obscure the way to escape routes and exits. The Regulations are concerned to:

(a) Provide a sufficient number and capacity of escape routes to a place of safety
(b) Protect escape routes from the effects of fire by enclosure, where necessary, and to limit the ingress of smoke
(c) Ensure the escape routes are adequately lit and exits suitably indicated.

The general principle of means of escape is that any person in a building confronted by an outbreak of fire can turn away from it and make a safe escape.

The number of escape routes and exits depends on the number of occupants in the room or storey, and the limits on travel distance to the nearest exit depend on the type of occupancy. The number of occupants in a room or storey is determined by the maximum number of people they are designed to hold, or calculated by using a floor space factor related to the type of accommodation which is used to determine occupancy related to floor area as set out in Approved Document B. The maximum number of occupants determines the number of escape routes and exits; where there are no more than 50 people one escape route is acceptable. Above that number, a minimum of 2 escape routes is necessary for up to 500 and up to 8 for 16000 occupants. Maximum travel distances to the nearest exit are related to purpose-group types of occupation and whether one or more escape routes are available. Distances for one direction escape are from 9.0 to 18.0 and for more than one direction escape from 18.0 to

45.0, depending on the purpose groups defined in Approved Document B.

Internal fire spread (linings) Fire may spread within a building over the surface of materials covering walls and ceilings. The Regulations prohibit the use of materials that encourage spread of flame across their surface when subject to intense radiant heat and those which give off appreciable heat when burning. Limits are set on the use of thermoplastic materials used in rooflights and lighting diffusers.

Internal fire spread (structure) As a measure of ability to withstand the effects of fire, the elements of a structure are given notional fire resistance times, in minutes, based on tests. Elements are tested for the ability to withstand the effects of fire in relation to:

(a) Resistance to collapse (loadbearing capacity) which applies to loadbearing elements
(b) Resistance to fire penetration (integrity) which applies to fire separating elements
(c) Resistance to the transfer of excessive heat (insulation) which applies to fire separating elements.

The notional fire resistance times, which depend on the size, height and use of the building, are chosen as being sufficient for the escape of occupants in the event of fire.

The requirements for the fire resistance of elements of a structure do not apply to:

(1) A structure that only supports a roof unless
  (a) the roof acts as a floor, e.g. car parking, or as a means of escape
  (b) the structure is essential for the stability of an external wall which needs to have fire resistance
(2) The lowest floor of the building.

Compartments To prevent rapid fire spread which could trap occupants, and to reduce the chances of fires growing large, it is necessary to subdivide buildings into compartments separated by walls and/or floors of fire-resisting construction. The degree of subdivision into compartments depends on:

(a) The use and fire load (contents) of the building
(b) The height of the floor of the top storey as a measure of ease of escape and the ability of fire services to be effective
(c) The availability of a sprinkler system which can slow the rate of growth of fire.

The necessary compartment walls and/or floors should be of solid construction sufficient to resist the penetration of fire for the stated notional period of time in minutes. The requirements for compartment walls and floors do not apply to single-storey buildings.

Concealed spaces Smoke and flame may spread through concealed spaces, such as voids above suspended ceilings, roof spaces and enclosed ducts and wall cavities in the construction of a building. To restrict the unseen spread of smoke and flames through such spaces, cavity barriers and stops should be fixed as a tight fitting barrier to the spread of smoke and flames.

External fire spread To limit the spread of fire between buildings, limits to the size of 'unprotected areas' of walls and also finishes to roofs, close to boundaries, are imposed by the Building Regulations. The term 'unprotected area' is used to include those parts of external walls that may contribute to the spread of fire between buildings. Windows are unprotected areas as glass offers negligible resistance to the spread of fire. The Regulations also limit the use of materials of roof coverings near a boundary that will not provide adequate protection against the spread of fire over their surfaces.

Access and facilities for the Fire Services Buildings should be designed and constructed so that:

- Internal firefighting facilities are easily accessible
- Access to the building is simple
- Vehicular access is straightforward
- The provision of fire mains is adequate.

Resistance to the passage of heat

The interior of buildings is heated by the transfer of heat from heaters and radiators to air (conduction), the circulation of heated air (convection) and the radiation of energy from heaters and radiators to surrounding colder surfaces (radiation). This internal heat is transferred to and through colder enclosing walls, roof and floors by conduction, convection and radiation to colder outside air. As long as the interior of buildings is heated to a temperature above that of outside air, transfer of heat from heat sources to outside air will continue. For the sake of economy in the use of expensive fuel and power sources, and to conserve limited supplies of fuel, it is sensible to seek to limit the rate of transfer of heat from inside to outside. Because of the variable complex of the modes of transfer of heat it is convenient to distinguish three separate modes of heat transfer as conduction, convection and radiation.

Conduction is the direct transmission of heat by contact between particles of matter, convection the transmission of heat by the motion (circulation) of heated gases and fluids, and radiation the transfer of heat from one body of radiant energy through space to another by a motion of vibration in space which radiates equally in all directions.

Conduction

The speed or rate at which heat is conducted through a material depends mainly on the density of the material. Dense metals conduct heat more rapidly than less dense gases. Metals have high and gases have low conductivity. Conductivity (k) is the rate of heat per unit area conducted through a slab of unit thickness per degree of temperature difference. It is expressed in watts per metre thickness of material per degree kelvin (W/mK) where W (watt) is the unit of power which is equivalent to joules (the unit of heat) per second (J/s) and the temperature is expressed in kelvin (K).

Convection

The density of air that is heated falls and the heated air rises and is replaced by cooler air. This, in turn, is heated and rises so that there is a continuing movement of air as heated air loses heat to surrounding cooler air and cooler surfaces of ceilings, walls and floors. Because the rate of transfer of heat from air to cooler surfaces varies from rapid transfer through thin sheets of glass in windows to an appreciably slower rate of transfer to insulated walls by conduction, and because of the variability of the exchange of cold outside air with warm inside air by ventilation, it is not possible to quantify heat transfer by convection. Usual practice is to make an assumption of likely total air changes per hour or volume (litres) per second depending on categories of activity in rooms and then to calculate the heat required to raise the temperature of the fresh, cooler air introduced by natural or mechanical ventilation, making an assumption of the temperature of inside and outside air.

Radiation

Energy from a heated body radiating equally in all directions is partly reflected and partly absorbed by another cooler body (with the absorbed energy converted to heat). The rate of emission and absorption of radiant energy depends on the temperature and the nature of the surface of the radiating and receiving bodies. The heat transfer by low temperature radiation from heaters and radiators is small whereas the very considerable radiant energy from the sun that may penetrate glass and that from high levels of artificial illumination is converted to appreciable heat inside buildings. An estimate of the solar heat gain and heat gain from artificial illumination may be assumed as part of the heat input to buildings and used in the calculation of heat input and loss.

Transmission of heat

The practical guidance in Approved Document L to meeting the requirements from Part L of Schedule 1 to the Building Regulations 1991, for the conservation of fuel and power, is mainly directed to limiting the loss of heat through the fabric (walls, floors and roofs) of buildings by establishing maximum values for the overall transmission of heat, the '*U*' value, through walls, roofs and floors and to limiting the size of glazed areas.

Because of the complexity of the combined modes of transfer of heat through the fabric of buildings it is convenient to use a coefficient of heat transmission, the *U* value. This air-to-air heat transmittance coefficient, the *U* value, takes account of the transfer of heat by conduction through solid materials and gases, convection of air in cavities and across inside and outside surfaces and radiation to and from surfaces. The *U* value is expressed as W/m^2K. A high *U* value indicates comparatively high rates of overall transmission and a low *U* value indicates low rates.

The maximum *U* values given in Approved Document L are 0.45 W/m^2K for exposed walls, floors, ground floors and roofs for buildings other than dwellings.

The loss of heat through windows and rooflights is

limited by setting maximum sizes related to floor and roof areas as:

Windows
35% of exposed wall area for places of assembly, offices and shops
15% of exposed wall area for industrial and storage.
Rooflights
20% of roof area for places of assembly, offices, shops, industrial and storage.

The methods of making an estimate of rates of heat transfer that are suggested in Approved Document L are the elemental approach and calculation procedures discussed below.

Elemental approach The elemental approach method is used to select the components of the construction of elements of the fabric of buildings that will not exceed stated maximum U values for walls, floors and roofs. Two methods of calculating what thickness of insulation may be required for an element of structure are set out in Approved Document L.

The first method depends on the use of tables giving basic thickness of a range of insulation materials required to achieve the maximum U values and allowable reduction in the basic thickness of insulation where a small range of conventional components are used in the construction, e.g. 100 brick outer skin, a cavity and 100 concrete block inner skin to a wall.

The second method takes account of the insulating properties of the whole construction of an element by calculating the thermal resistance of each component of the element and the standard values for the resistance of air spaces and surfaces to transmission of heat. It may be used to make a more accurate calculation of U values, particularly where the materials of the components are not included in the tables used in the first method.

Resistivity, which is the reciprocal of conductivity, expresses the resistance of a material to the transfer of heat. Resistivity multiplied by the thickness of a material is defined as resistance which may be determined by dividing the thickness of a material, in metres, by its thermal conductivity.

Calculation procedures There are two calculation procedures. The first calculation procedure is an alternative to the elemental approach by allowing variations in the level of insulation of elements and the areas of windows and rooflights to facilitate the use of other areas of glass and rooflights than those set out above, as long as the rate of heat loss would be no greater than that determined by the elemental approach. The second calculation procedure allows greater freedom of selection of components of the elements of structure by reference to an energy target that can take account of notional internal temperatures, internal heat gains from people, cooking, hot water systems, lighting and solar heat gain. The energy use calculated by this procedure should be no greater than it would be for a similar building that complies with the elemental approach.

Condensation

During recent years increased expectation of thermal comfort in buildings and the need to conserve limited supplies of fuel and power has led to improved levels of insulation in the fabric of buildings and the common use of weatherseals to opening windows that has restricted natural ventilation. These changes have led to the likelihood of increased levels of humid conditions that cause condensation on the inner faces of cold surfaces such as glass in windows and the inside of thin metal sheet weathering.

The limited capacity of air to take up water in the form of water vapour increases with temperature so that the warmer the air, the greater the amount of water vapour it can hold. The amount of water vapour held in air is expressed as a ratio of the actual amount of water vapour in the air to the maximum which the air could contain at a given temperature. This relative humidity (rh) is given as a percentage. Air is saturated at 100% relative humidity and the temperature at which this occurs is defined as the dew point temperature. When the temperature of warm moist air falls to a temperature at which its moisture vapour content exceeds the saturation point, the excess moisture vapour will be deposited as water. This will occur, for example, where warm moist air comes into contact with cold window glass, its temperature at the point of contact with the glass falls below that of its saturation point and the excess moisture vapour forms as droplets of water on the inside window surface as condensation. Thus, the greater the amount of moisture vapour held in the air and the greater the temperature difference between the warm inside air and the cold window surface, the more the condensation.

The main sources of moisture vapour in air in buildings are moisture given off by occupants, moisture given off from cooking and flueless heaters, bathing, clothes washing and drying, moisture generating processes and the drying out from a new building.

Internal temperatures which are high relative to cold outside air will tend to produce high levels of condensation on cold surfaces from high levels of moisture vapour. An atmosphere that contains high levels of moisture vapour is said to be humid. The level of humidity that is acceptable for comfort in buildings varies from about 30% to 70% relative humidity. Low levels of humidity, below about 20%, may cause complaints of dry throats and cause woodwork to shrink and crack. High levels of humidity, e.g. above 70%, may cause discomfort and lead to condensation on cold surfaces, mould growth and excess heat. For comfort in buildings and to limit the build-up of humidity it is necessary to provide a degree of ventilation for an adequate supply of oxygen and to limit fumes, body odour and smells and to exchange drier fresh air from outside with humid, stale inside air. The level of ventilation required depends on the activity and number of people in a given space and sources of heat and water that will contribute to increased humidity.

The position of the layer of insulation in the thickness of roof and wall construction is determined mainly by convenience in fixing the material as, e.g. over and across purlins and sheeting rails in single-storey framed structures. The position of the insulation will affect the temperature of those parts of the construction relative to inside and outside temperatures and the possibility of condensation of warm moist air on cold surfaces. To minimise the possibility of condensation the construction, where practical, should be on the inside of the insulation where it will be maintained at inside air temperature. This arrangement, which is sometimes referred to as warm construction, is illustrated in Fig. 54 where the whole of the construction including the weathering membrane is below the insulation in a roof. This unusual arrangement is adopted to protect the weathering membrane from the extremes of temperature variations that can occur between day and night.

More usually the insulation is laid under the weathering layer on the deck of both solid concrete and profiled steel decking, as illustrated in Fig. 54, showing typical warm construction or warm roofs. Where the insulation is on or towards the inside face, as for example with ceiling insulation, the construction will be cold or cold roof construction.

With profiled metal and fibre cement sheeting it is usual practice to lay the insulation across the purlins and sheeting rails under the roof and wall covering. The covering will, therefore, be at outside air temperature and, as such, will be a cold construction component of roofs and walls and be liable to suffer condensation of warm moist inside air that penetrates the insulation and lining. The likelihood of condensation on the cold

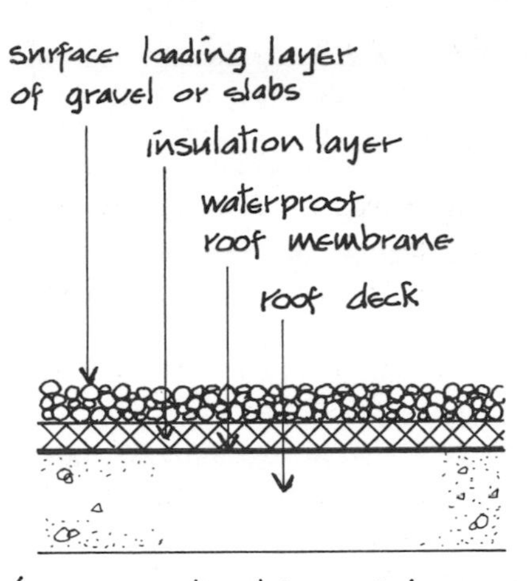

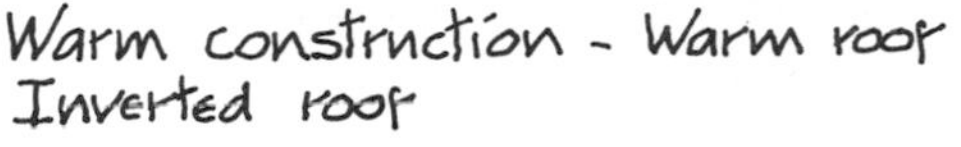

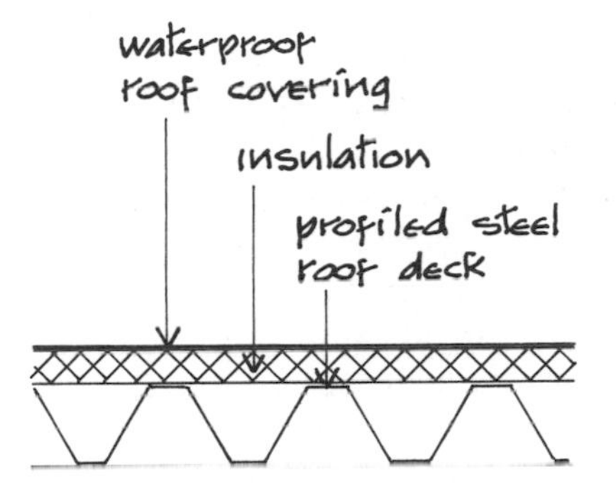

Warm construction - Warm roof

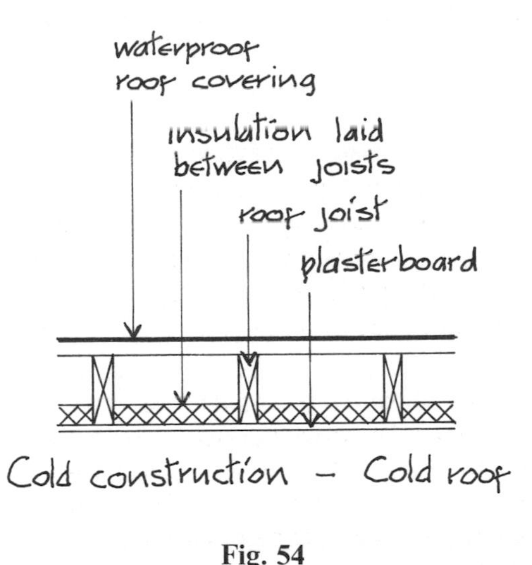

Fig. 54

under surface of sheeting depends on the humidity and temperature of the inside air, the extent to which warm moist air can penetrate the insulation and lining sheets and the relative temperature of inside and outside air. The humidity of the inside air depends on temperature and the activities inside the building. The air inside a laundry will be much more humid than that inside a dry heated store and much more likely to cause condensation on cold sheeting.

Heavy condensation on the underside of steel sheeting may, in time, cause corrosion of steel and saturation of porous insulating materials which will lose their effectiveness as insulation. Condensation on the underside of cold fibre cement sheets will have no adverse effect on the material.

With comparatively dry inside air it is generally sufficient to accept that the insulation and any supporting lining sheets will serve as an adequate check against appreciable volumes of inside air causing condensation on the underside of cladding.

With medium and high levels of internal humidity it is generally necessary to provide some form of check or barrier to the penetration and build-up of warm moist air that might otherwise condense on cold cladding. The two ways of minimising the build-up of warm moist air under sheeting are by ventilating the space between the sheeting and the top of the insulation or by making some form of check or barrier to the penetration of warm moist air.

Ventilation

Ventilation of the space between sheeting and insulation combined with the check to penetration of inside air provided by the insulation itself and any under lining sheets used, is the most straightforward way of minimising condensation. The space between the sheeting and the insulation is ventilated by providing spacers laid over the insulation, under the troughs of the profiled sheeting and over the purlins and sheeting rails to provide a space of 100 which is ventilated through the open, unsealed joints between sheets to outside air. This passive ventilation is generally sufficient to prevent an excessive build-up of warm moist air under the sheeting.

With medium and high levels of humidity in buildings there is a pressure drive of internal moisture vapour towards colder outside air. Because of the vapour pressure drive, moisture vapour may penetrate joints between insulation and lining panels and also, to an extent, penetrate the insulation itself. With low to medium levels of internal humidity, foamed insulation boards closely butted together and any internal lining panels will generally serve as a check to moisture vapour penetration. Where open textured insulation such as mineral fibre quilt or mat is used, there may be appreciable penetration of moisture vapour through the insulation. A ventilated space between the top of the insulation and the cladding will generally minimise the build-up of moisture vapour. Some moisture may, however, condense to water on the cold underside of the metal sheeting. To prevent condensation water falling on the insulation and saturing it, a breather paper membrane is stretched over the insulation and spacer bars. The micro porous holes in the breather paper allow moisture vapour to penetrate from below whilst retaining condensation water that will either evaporate or run down to eaves. The breather paper will, in addition, prevent cold ventilation air penetrating the open textured insulation thus reducing its insulating properties.

Glass fibre tissue faced foamed insulation boards laid over purlins, with the joints between the insulation boards taped on the top surface of the boards, will provide a sufficient check to the penetration of moisture vapour where there are low to medium levels of internal humidity.

Site-assembled and factory-made composite panels of profiled foamed or high density rock wool insulation that fits closely between profiled sheeting and profiled inner lining sheets with end and side laps between panels sealed will act as an effective check to the penetration of all but the highest levels of internal humidity.

Composite panels of insulation, closely fitting between sheeting and inner lining panels effectively sealed, and other forms of over purlin insulation to buildings with low levels of humidity may well not require ventilation of roof spaces.

The requirements for ventilation are illustrated in Fig. 55.

Resistance to the passage of sound

Where it is a requirement that the fabric of a building resists the pentration of sound it is necessary to ensure that there are as few openings as practical through which airborne sound may penetrate. Open windows, ventilators and small construction gaps will allow appreciable transfer of sound. Effective air seals

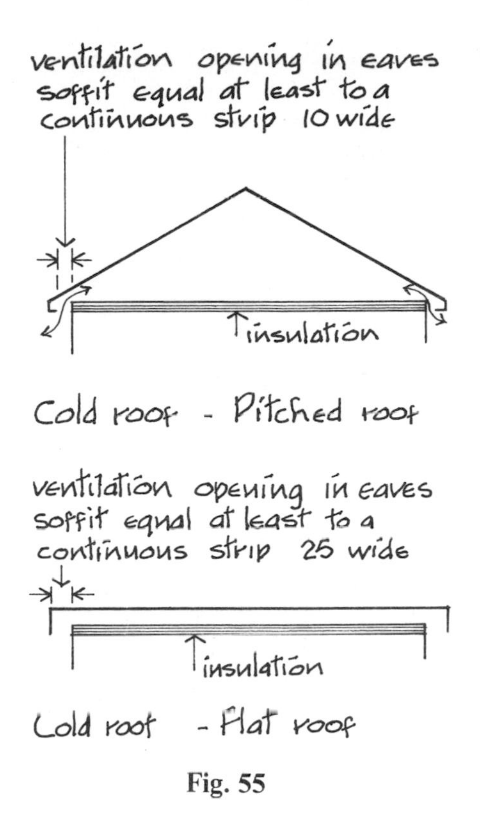

Fig. 55

around the opening parts of windows and doors, dampers to ventilators and careful sealing of construction gaps will appreciably reduce sound penetration. The thin metal skin of profiled steel sheeting affords no appreciable resistance to sound penetration, whereas a 60 or 80 thickness of mineral wool thermal insulation with a 20 air gap will provide appreciable reduction of low levels of sound.

Where effective resistance to high levels of sound is necessary as, for example, in buildings close to airports and where the sound from noisy industrial processes inside buildings has to be contained, it may be advantageous to adopt some form of solid enclosure such as brick or concrete to provide the required sound reduction.

Security

Many single-storey framed buildings which are occupied during working hours and days are vulnerable to damage by vandalism or forced entry, unless adequately guarded at night and weekends. Apart from the obvious risk of forced entry through windows, doors and rooflights there is a risk of entry by prising thin profiled sheeting from its fixing and so making an opening large enough for entry.

PROFILED SHEET COVERINGS FOR ROOFS AND WALLS

Corrugated iron sheets were first produced about 1830 from wrought or puddled iron which was either pressed or rolled to the corrugated profile. Because of the purity of the iron and the thickness of the sheets, the original corrugated iron sheets had a useful life of more than 50 years in all but the most corrosive atmospheres.

After steel making became a commercial proposition in about 1860, thin corrugated steel sheets were first used in Britain about 1880. These thin corrugated steel sheets were considerably cheaper and thinner than the original iron sheets but had a useful life of only about a quarter of that of the heavier iron sheets because the 'impurities' in the steel promoted more rapid corrosion even when the sheets were zinc coated. The comparatively rapid destructive corrosion of steel sheets led to the use of the non-corrosive alternative, asbestos sheets, early in the twentieth century.

In the early days of the use of corrugated steel sheets it was common practice to protect the sheets against corrosion by coating them with tar, pitch or paint. As long as these coatings adhered strongly to the whole of the surface and edges of the sheets, corrosion did not occur. Once the coating was broken, rust spread rapidly between the steel and the coating and frequent chipping and wire brushing of the exposed steel was necessary, followed by the application of protective coatings.

In 1883 zinc-galvanised corrugated steel sheets were imported into this country from Scandinavia and shortly afterwards similar sheets were produced here.

Galvanised, corrugated sheets are coated all over with a thin film of zinc, applied by dipping or spraying. Providing the zinc coating is sufficiently thick and adheres strongly to the steel below, it will inhibit rust for some years. On exposure the zinc coating first oxidises and then is converted to basic zinc carbonate which is largely insoluble in clean atmospheres and protects the zinc from further corrosion. In time, however, this surface film is worn away by the scouring action of wind and rain and progressive corrosion occurs until there is no further protection of the steel which rapidly corrodes.

The principal disadvantage of corrugated steel sheets is the difficulty of preventing corrosion of the steel for more than a few years. Other disadvantages are low thermal resistance, poor fire resistance and the poor appearance of the sheets due to the shallow corrugations.

Corrugated steel sheets coated with bitumen were first used in this country in 1923. During manufacture these sheets were immersed in hot bitumen which adhered strongly to the sheets and was effective in preventing corrosion of the sheets for some time. However, the bitumen coating by itself was highly inflammable and some spectacular fires, during which the bitumen burnt fiercely and the sheeting collapsed, caused this early type of protected sheeting to be withdrawn. A later, improved protective coating consisted of hot bitumen covered with a coating of asbestos felt bonded to the bitumen, which was in turn coated with bitumen. The combination of bitumen and asbestos felt protected the steel sheets from corrosion, improved thermal resistance and provided a fair degree of resistance to damage by fire. These sheets had a singularly unattractive appearance.

With improvements in the techniques of cold roll forming steel strip, a range of trapezoidal section profiled steel sheets was introduced about 1960. The improved strength and rigidity of these profiles allowed the use of thinner strip and wider spacing of supports than is possible with the shallow depth of corrugated sheet, and the bold angular trapezoidal profile contrasted favourably with the small, somewhat indeterminate profile of corrugated sheets. These trapezoidal profile sheets were at first coated with zinc, asbestos felt and bitumen and later with organic plastic coatings.

Plastic-coated profiled steel sheeting is the principal sheet material used for weather protection of single-storey framed buildings today. These sheets are made in a wide range of trapezoidal profiles and colours. Trapezoidal profile steel sheets are also the principal material used for roof decking where the sheets serve as support for insulation and a weather coat of built-up bitumen felt and thin membrane coatings for roofing.

Corrugated asbestos cement sheets were first used as an alternative to steel sheets because they are corrosion free and are comparable in cost to steel sheeting. The disadvantages of asbestos cement as a roof covering are its dull, cement grey colour, the thickness of the sheet which makes it impossible to form a close joint at overlaps of sheets for low pitched roofs and the brittle nature of the material which is readily fractured during handling, fixing and in use. Due to the hazards to health in the manufacture of asbestos products, these sheets are now made as fibre cement sheets.

PROFILED STEEL SHEETING

The advantages of steel as a material for roof and wall sheeting are that its favourable strength-to-weight ratio and ductility make it both practical and economic to use comparatively thin, lightweight sheets that can be cold roll formed to profiles with adequate strength and stiffness for handling and to support the loads normal to roof and wall coverings. The disadvantage of steel as a sheeting material is that it suffers rapid, progressive, destructive corrosion.

Protective coatings for profiled steel sheets

Corrosion of steel When exposed to air and water or salt solutions, steel undergoes a complex electrochemical change which is essentially a process of oxidation, termed corrosion, where the metal tends to return to an oxidised condition similar in composition to an iron ore from which it was produced. The corrosive oxidation of steel produces a reddish deposit on the surface of the steel, known as rust. The initial deposit of rust does not generally prevent further corrosion of the steel below so that progressive corrosion of the metal occurs. Because rust has poor strength and negligible ductility, the characteristics for which steel is used, corrosion reduces the usefulness of steel.

The corrosive process is a complex electrochemical action that depends on the characteristics of the metal, atmosphere and temperature. Corrosion of steel is most destructive in conditions of persistent moisture, atmospheric pollution and where different metals are in contact.

Zinc coating The most economic and effective way of protecting steel against corrosion is by coating it with zinc which corrodes more slowly than steel. The zinc coating acts as a barrier against contact of steel with the atmosphere and acts sacrificially to protect the steel at cut edges by galvanic action where two dissimilar metals are in contact in the presence of moisture and corrosion of only one takes place. The most reactive of the two will become the anode in a natural electric cell and will oxidise to protect the cathode from corrosion. Thus the zinc coat protects the steel as it is anodic to steel in a galvanic action and corrodes sacrificially to the benefit of the steel.

Hot-dip galvanising This is the most commonly used method of applying a zinc coating to steel. There are three stages to the process. The steel is first degreased and cleaned in cold, dilute hydrochloric acid, the cleaned steel is given a prefluxing treatment and then fully immersed in molten zinc at a temperature of about 450°C. The thickness of the coating depends on the time of immersion, withdrawal rate and the temperature of the molten zinc. The usual coatings of zinc are a minimum of 275 g/m² including both sides for cladding. The zinc coating adheres strongly to the steel through the formation of a very thin alloy layer, between the steel and the zinc, which bonds strongly to both metals. The protection afforded by the zinc coating depends on the thickness of the coating and atmospheric conditions. The products of the corrosion of zinc in rural areas are of low solubility that tend to inhibit corrosion of the zinc below for periods of up to 30 years, whereas the products of corrosion in industrial areas are soluble and afford less protection for periods of about 7 years. The life expectancy of a protective coating in a particular atmosphere is proportional to the weight of the zinc coating.

In rural, dry sub-tropical and marine tropical areas where the rate of atmospheric corrosion is low, hot-dip galvanised zinc coating will provide by itself adequate protection of steel sheets where the metallic grey colour of the sheets provides an acceptable finish to buildings.

Zalutite Zalutite is a protective coating of an alloy of zinc, aluminium and silicon which is about twice as durable as a zinc coating. It is used as a protective coating by itself as a low cost form of protection where the appearance of the surface is not important. The usual thickness of this coating is 185 g/m² including both sides.

Organically (plastic) coated profiled steel sheets The majority of profiled steel sheets used today are organically coated with one of the plastic coatings available for the protection afforded by the plastic finish and the range of colours available. Plastic coatings to galvanised zinc coated steel sheets serve as a barrier to atmospheric corrosion of zinc, the erosive effect of wind and rain and protection from damage during handling, fixing and in use. Also they serve as a means of applying a colour to the surface of the sheets. The principal advantage of the extra cost of the plastic coating is in the application of a colour to what is otherwise a drab metallic grey zinc coating.

Colour is applied to organically coated steel sheets by the addition of pigment to the coating material. The effect of ultraviolet radiation and the weathering effect of wind and rain is to gradually bleach the colour pigment in the coating. Loss of colour is not uniform over the whole surface of profiled sheets, it being most pronounced on south-facing slopes and sides of buildings and irregular on the ridges and flanges of the sheets. This varied loss of colour over a number of years spoils the appearance of buildings.

Organically coated steel sheets are used principally for the benefit of the colours available so that the inevitable loss of colour may in time become unacceptable from the point of view of appearance. The term 'life expectancy to first maintenance' used in relation to coloured, organically coated steel sheets, expresses the term in years that a particular coating will adequately retain its colour to a comparatively stringent standard of appearance before overpainting is deemed necessary. The term 'life expectancy' to first maintenance is not generally used to define the useful life of the coating as protection against corrosion and damage, which may be considerably longer than that of colour retention.

Organic coatings for profiled steel sheets

uPVC-polyvinyl chloride This is the cheapest and most used of the organic plastic coatings and is known as 'Plastisol'. The comparatively thick (200 microns) coating that is applied over a zinc coating provides good resistance to damage in handling, fixing and in use and good resistance to normal weathering agents. The material is ultraviolet stabilised to retard degradation by ultraviolet light and the consequent chalking and loss of colour. The durability of the coating is good as a protection for the zinc coating below but the life expectancy to first maintenance of acceptable colour retention is only of the order of 10 to 20 years.

Polyvinyl chloride is an economic, tough, durable, scratch-resistant coating that will provide good protection of the zinc coating and steel below for many years but has poor colour retention.

Acrylic-polymethyl methacrylate – PMMA This organic plastic, which is about twice the price of uPVC, is applied with heat under pressure as a laminate to galvanised zinc steel strip to a thickness of 75 microns. It forms a tough finish with high strength, good impact resistance and good resistance to damage by handling, fixing and in use. It has excellent chemical resistance and its good resistance to ultraviolet radiation gives a life expectancy of acceptable colour retention to first maintenance of up to 20 years. The hard smooth finish of this coating is particularly free from dirt staining.

PVF-polyvinylidene fluoride A comparatively expensive organic plastic coating for profiled steel sheets which is used as a thin (25 microns) coating to zinc coated steel strip for its excellent resistance to weathering, excellent chemical resistance, durability and resistance to all high energy radiation. Because this coating is thin it may be damaged by careless handling and fixing. The durability of this coating is good as protection for the zinc coating and the steel sheet and its life expectancy to first maintenance in relation to colour retention is better than 15 years and up to 30 years.

Silicone polyester This is the cheapest of the organic coatings used for galvanised steel sheet. It is suitable for use in temperate climates where life expectancy to first maintenance is from five to seven years. It is not suitable for use in marine and hot humid atmospheres or where there is aggressive industrial pollution of the atmosphere. The galvanised steel sheets are primed and coated with stoved silicone polyester to a thickness of 25 microns. The coating provides reasonable protection against damage in handling, fixing and use, good resistance to ultraviolet radiation and a life expectancy to first maintenance of five to seven years.

Bitumen and fibre mineral coatings Galvanised steel strip is coated with hot bitumen, a second layer of a composite of mineral fibre and bitumen, finished with a coloured alkyd resin. This coating provides good protection against corrosion, moderate protection against damage by handling, fixing and in use and a life expectancy of 10 to 15 years to first maintenance.

Insulation materials

Materials used as insulation may be grouped as inorganic and organic insulants. Inorganic insulants are made from naturally occurring materials that are formed into fibre, powder or cellular structures that have a high void content, e.g. glass fibre and mineral wool (rockwool). Inorganic insulants are generally incombustible, do not support spread of flame, are rot and vermin proof, are permeable to moisture vapour and generally have a higher *U* value than organic insulants.

The inorganic insulants most used as insulation to steel sheeting are glass fibre and rockwool in the form of loose fibres, mats and rolls of felted fibres and semi-rigid boards, batts and slabs of compressed fibres.

Organic insulants are based on hydrocarbon polymers in the form of thermosetting or thermoplastic resins to form structures with a high void content, e.g. polystyrene, polyurethane, isocyanurate and phenolic. Organic insulants generally have a lower *U* value than inorganic insulants, are combustible, and support the spread of flames more readily than inorganic insulants. Thermoplastic insulants have comparatively low softening points.

The organic insulants most used as insulation to steel sheeting are expanded polystyrene in the form of boards, extruded polystyrene in the form of boards and polyurethane and isocyanurate in the form of preformed rigid boards.

The cheapest materials used for insulation are glass fibre, rockwool and polystyrene.

Table 1 gives *U* values for insulating materials.

Table 1. Insulating materials

Material	*U* value W/m^2K
Glass fibre	0.04
Rockwool	0.037
EPS	0.04
XPS	0.025
PIR	0.02
PUR	0.022

EPS = expanded polystyrene
XPS = extruded polystyrene
PIR = rigid polyisocyanurate
PUR = rigid polyurethane.

Where there is appreciable internal humidity in buildings it is necessary to prevent warm, moist air condensing to water on cold metal surfaces such as sheeting and spacers and so causing corrosion. Condensation may be prevented or minimised by ventilation of roof and wall cavities under sheeting or by the use of a vapour check or barrier on the warm side of the construction. Steel inner lining panels and preformed organic insulants serve as checks to the penetration of medium levels of internal humidity.

The practical guidance in Approved Document L to meeting the requirement from Part L of Schedule 1 to the Building Regulations 1991, for the conservation of fuel and power, gives a maximum *U* value for the walls and roofs of buildings, other than dwellings, of 0.45 W/m^2K. As thin steel sheets provide negligible resistance to the passage of heat, insulation is used to provide the major part of resistance to the transfer of heat. An insulation thickness of 52 with a material of 0.025 conductivity and 73 with a material of 0.03

conductivity is necessary to achieve a U value of 0.45 W/m^2K.

Profiled steel cladding

The term cladding is a general description for materials used to clothe or clad the external faces of buildings to provide protection against wind and rain. Profiled steel sheeting is one form of cladding. Steel sheet cladding for roofs is known as roof sheeting or roof cladding, and cladding for walls as wall sheeting, walling or sidewalling. Because cladding serves as a weathering skin, insulation has to be laid or fixed under it as illustrated in Fig. 56.

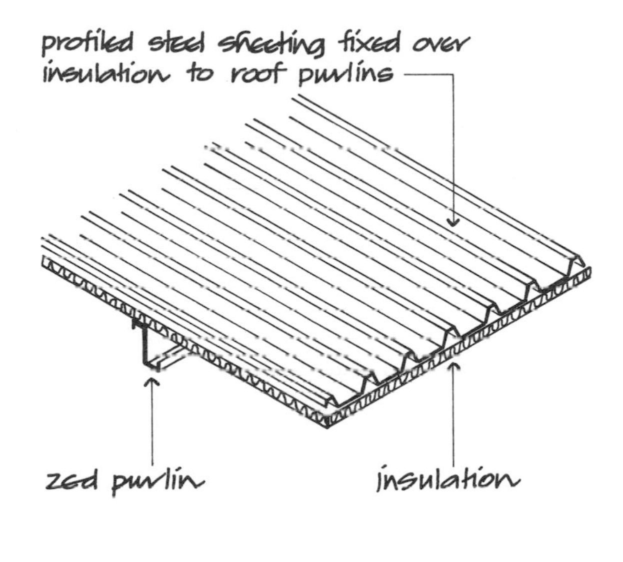

Fig. 56

Corrugated steel sheeting, illustrated in Fig. 57, has shallow corrugations 19 deep that provide longitudinal strength and rigidity sufficient for the limited centres of support common in small buildings and economy in the reduction of the steel strip by forming. The cover width of corrugated sheet is 914 as compared to 860 for a trapezoidal profiled sheet 48 deep.

The typical trapezoidal profile steel sheet is formed with trapezoidal section ridges with flat lower flanges as illustrated in Fig. 58. The depth of the ridges provides longitudinal strength and stiffness for support of dead and imposed loads. The thin flat bottom flange of the sheet between ridges is subject to buckling in handling, fixing and local loading, such as wind uplift, to the extent that it may be necessary to improve stiffness with shallow longitudinal ribs in both the wide lower flanges and the top of ribs as illustrated in Figs. 57 and 58.

During recent years standing seam profiled sheets

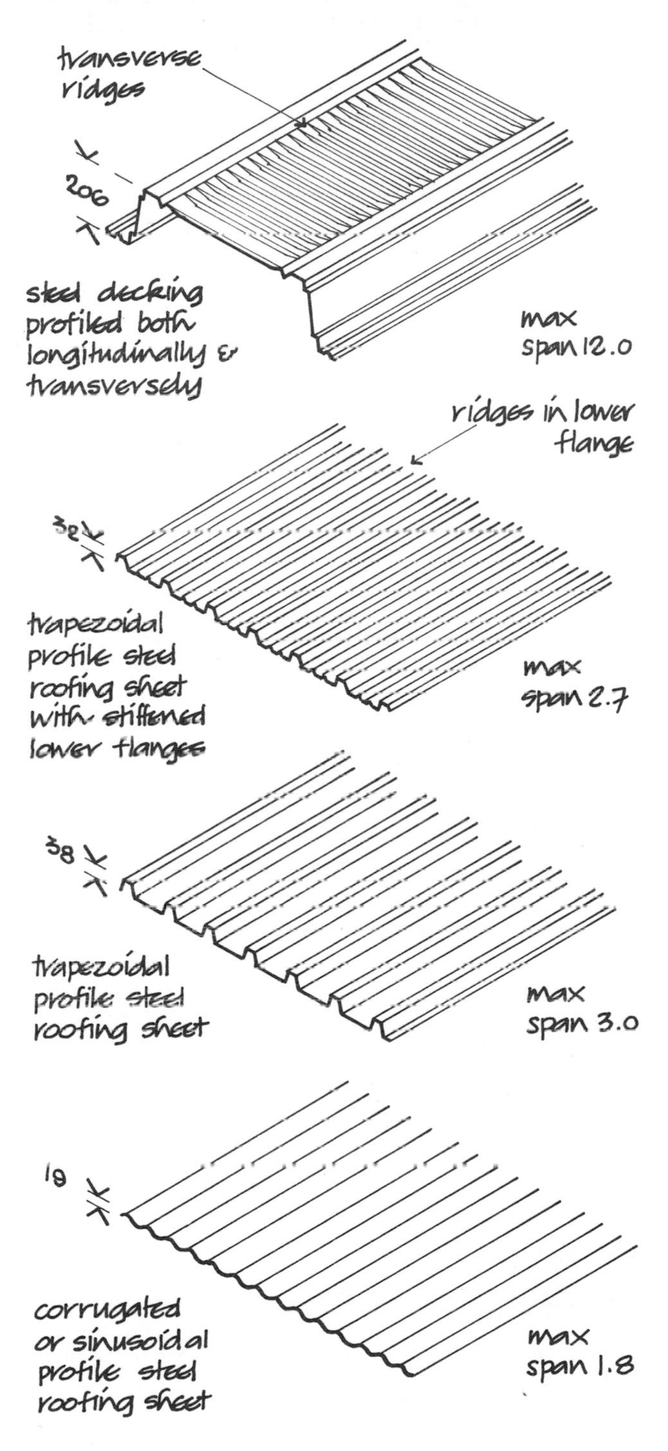

Fig. 57

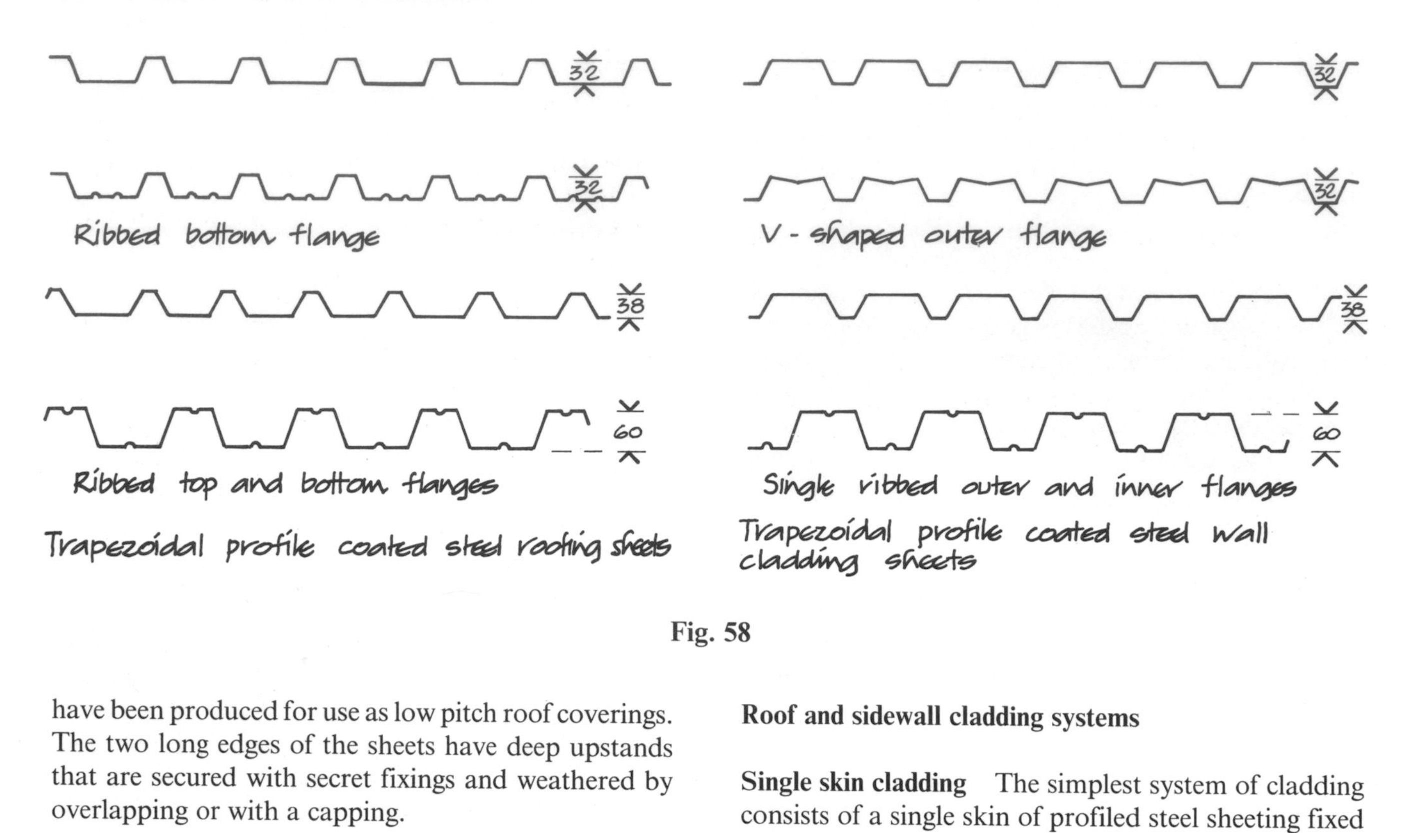

Fig. 58

have been produced for use as low pitch roof coverings. The two long edges of the sheets have deep upstands that are secured with secret fixings and weathered by overlapping or with a capping.

Roof and sidewall cladding systems

Single skin cladding The simplest system of cladding consists of a single skin of profiled steel sheeting fixed

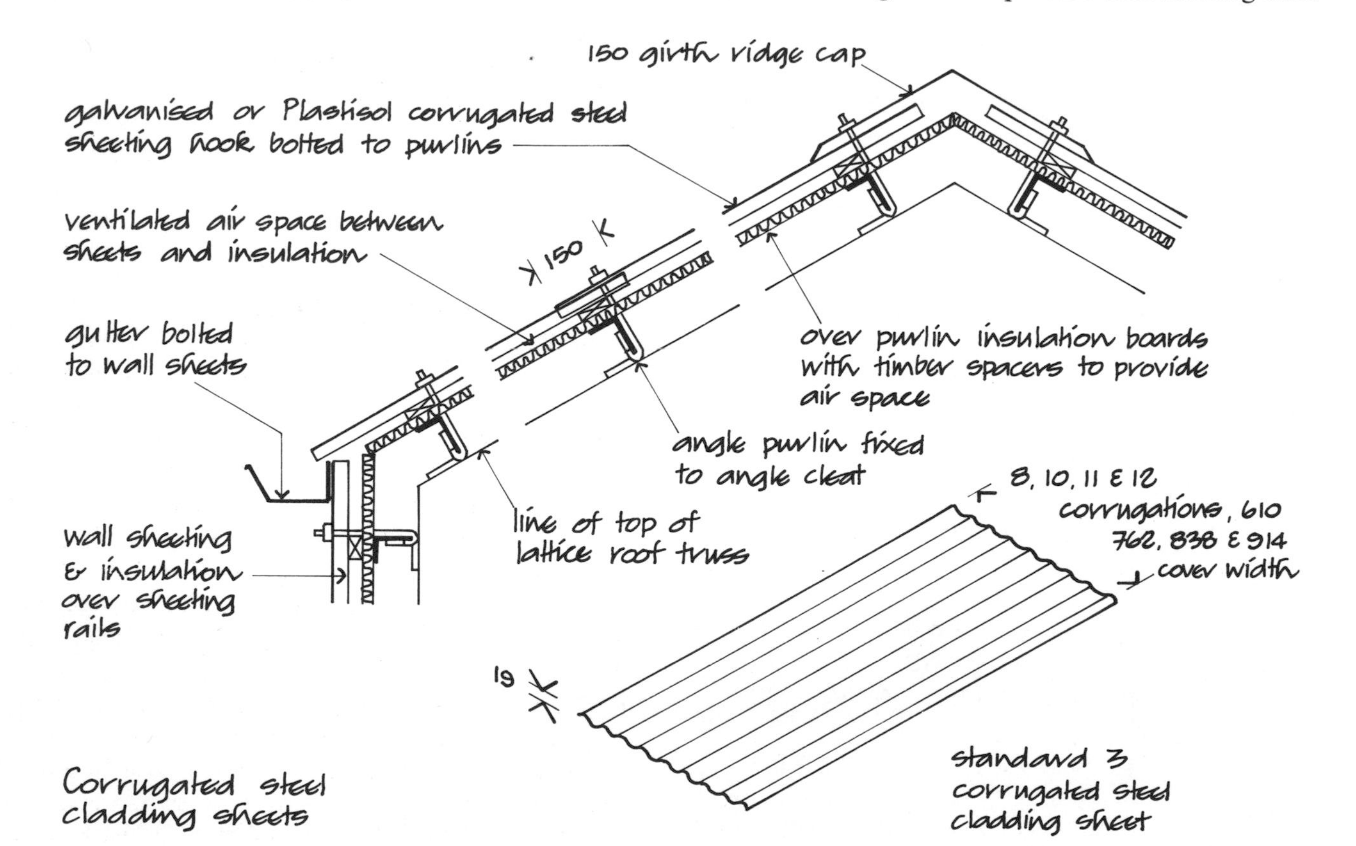

Fig. 59

directly to purlins and sheeting rails without insulation or inner lining. This, the cheapest form of cladding, is used for buildings that do not require heating such as stores, warehouses, sheds and buildings heated by their own processes, e.g. foundries. Corrugated or trapezoidal sheeting is fixed directly to roof purlins and sheeting rails.

Over purlin insulation without lining The most straightforward and economic system of supporting insulation under cladding is to use semi-rigid or rigid preformed boards laid across roof purlins and sheeting rails as illustrated in Fig. 59. Where corrugated or shallow depth trapezoidal sheeting is used the necessarily closely spaced purlins and sheeting rails will provide adequate support for the insulation boards without the need for inner lining sheets. Where the purlins and sheeting rails are more widely spaced the insulation boards are supported by T-section bars laid across the purlins at intervals suited to the support of the boards. To provide an air space for passive ventilation between cladding and insulation, to minimise condensation, timber spacers are used as illustrated. Corrugated cladding is secured with galvanised hook bolts that are bolted through the ridges of the cladding and hooked to angle purlin, sheeting rails and trapezoidal sheeting by screws. This system of cladding is suitable for buildings with low to medium levels of internal humidity where the appearance of insulation boards as a soffit and inner finish to walls is acceptable.

Over purlin insulation with inner lining Where mineral fibre mat insulation is used and where more rigid forms of insulation will not be self-supporting between widely spaced purlins, it is necessary to use profiled inner lining sheets or panels to provide support for insulation.

Lining sheets or panels are cold, roll formed, steel strips with shallow depth profiles adequate to support the weight of insulation. The sheets of panels are hot dip galvanised and may be coated on the exposed side with a heat cured, polyester coating or lining enamel as a protective and decorative coating.

The most commonly used system of over purlin insulation with lining panels consists of cladding sheets over mineral fibre mat or quilt laid on lining panels. To prevent compression of the loose mat or quilt, its thickness is maintained by spacers fixed between cladding and lining panels as illustrated in Fig. 60. The spacers are usually of Zed section cold formed steel of the same or greater depth than the insulation. The spacers are screwed through ferrules, through the troughs of the lining panels to the roof purlins and sheeting rails. The roof and walling sheets are then fixed by screwing through the crown or troughs of the sheeting, to the spacers, with self-tapping screws with neoprene washers. In this system the plastic ferrules serve as a thermal break. The space between the top of the sheeting and the insulation is passively ventilated to minimise condensation, through the open ends and side laps of the sheeting, where there are low or medium levels of humidity in the building. Where the internal humidity is high, with low pitch roofs and in exposed positions, a breather paper is often spread over the insulation and the top of spacers under sheeting which has sealed end and side laps to exclude rain, and with the roof and wall cavity below the sheets sealed. The breather paper, which is impermeable to water, protects the insulation from any rain or water condensation yet allows moisture vapour to penetrate it.

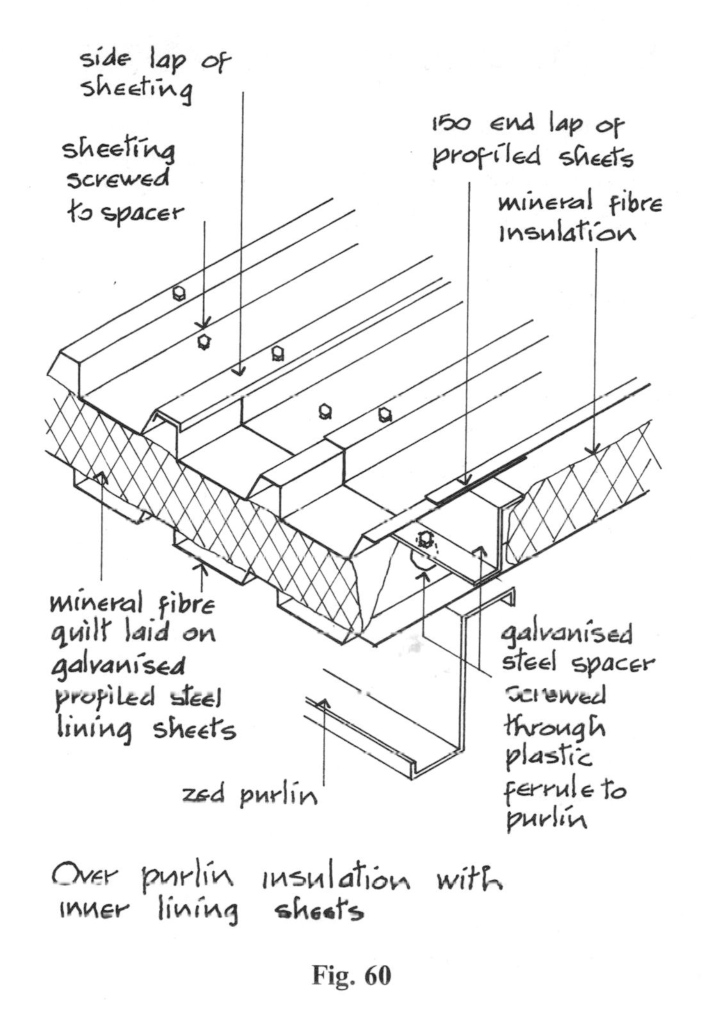

Fig. 60

Over purlin composite insulation and sheeting (site assembled) This system comprises a core of rigid preformed lightweight insulation shaped to match the profile of both the roof sheeting and the inner lining sheets. The advantages of this system are that the insulation is sufficiently rigid for use without spacers between the roof, wall and lining sheets, the insulation fills and seals the space between the sheets and that insulants with a low *U* value such as PUR and EPS can be used. The separate roof and wall sheeting, preformed insulation and inner lining sheets are assembled on site and fixed directly to the purlins and rails with self-tapping screws, driven through the troughs or crowns of the profiles as illustrated in Fig. 61. The side and end laps between lining sheets are sealed against the penetration of moisture vapour. Coated steel eaves and ridge filler pieces are used to seal the insulation cavity. This system of cladding, which is somewhat more expensive than the fibre quilt and spacers system, has the advantage of no thermal bridge due to spacers, continuity of insulation and the use of efficient insulants for cladding buildings with medium to high levels of internal humidity. The disadvantage is that the intimate contact of the insulation with the underside of the sheeting may lead to a considerable build-up of heating of sheets due to solar radiation and so reduce the useful life of coated sheeting.

Over purlin composite (factory formed) A more expensive system of composite panels consists of

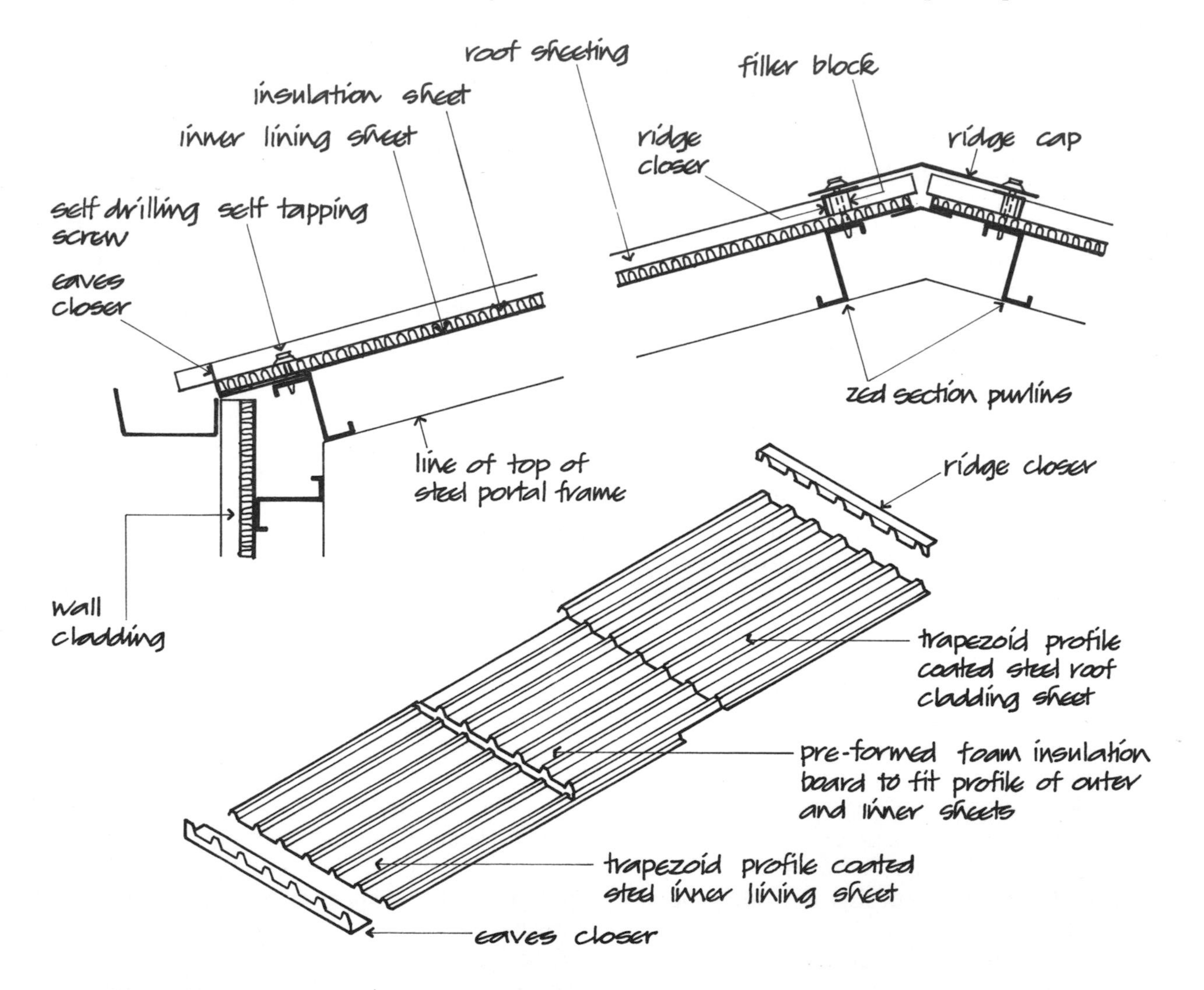

Fig. 61

factory-formed panels with a foamed insulation core enclosed and sealed by profiled sheeting and either flat or profiled inner lining panels made for batten capping for low pitch roofs. The advantages of the composite panels are that the two panels and their insulating core act together structurally to improve loading characteristics and that the panels have secret fixings for appearance. A particular disadvantage of these panels is that the sealed edges of the sheets may act as a very narrow thermal bridge. Because the insulation is sealed in the panels there is no need for ventilation of roof spaces.

The insulating core is usually of foamed insulation such as PUR. Figure 62 is an illustration of factory-formed panels.

Standing seams Standing seams are principally used for low and very low pitch roofs to provide a deep upstand as weathering to the side joints of sheeting and also for the benefit of secret fixings for appearance. To avoid the complication of detail at end laps with standing seams, these sheets are usually provided in lengths that can span from ridge to eaves. The standing seams for sheets that are assembled on site over insulation and inner lining panels are made as interlocking standing seams. With this system the roof and wall sheeting may be the same. Figure 63 is an illustration of standing seam cladding with secret fixings between panels, screwed to spacers and hooked to upstands of the standing seams. The spacers serve to maintain the space between roof sheeting and lining sheets for mineral fibre quilt insulation.

An advantage of the secret fixings used with standing seams is that the clip on fixing allows some freedom for the thermal movement of long sheets that might otherwise be deformed if thermal movement were restrained by screwed fixings.

Because of the considerable stiffening effect of the depth of the standing seams, the depth of the profiles of the sheet can be reduced.

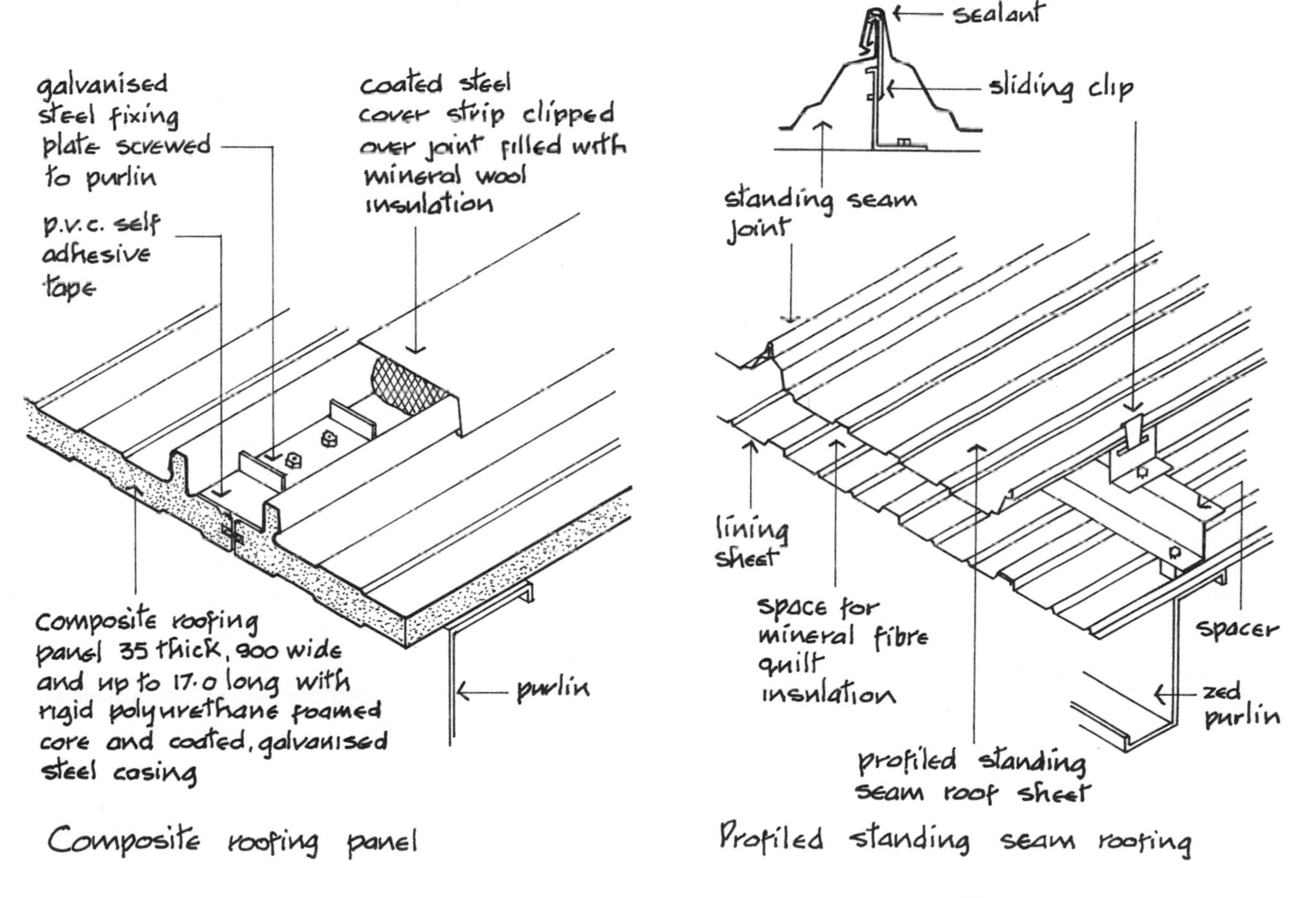

Fig. 62

Fig. 63

Fasteners Steel cladding, lining sheets and spacers are usually fixed with coated steel or stainless steel self-tapping screws illustrated in Fig. 64. The screws are mechanically driven through the sheets into purlins or spacers. These primary fasteners for roof and wall sheeting may have heads coloured to match the colour of the sheeting.

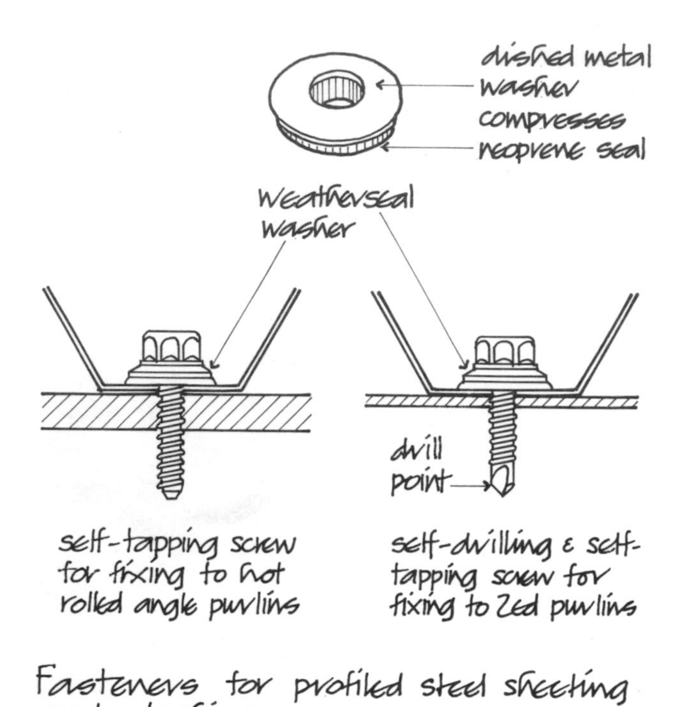

Fig. 64

For appearance the roof and wall sheeting screws are fixed in the trough of the sheets, where they are least visible. Correctly driven home, the neoprene washer under the head of the screws should make a watertight joint. Recently it has become common to fix roof sheeting through the crown of the profile, particularly with low and very low pitch roofs, to minimise the potential for damage and penetration of water through poorly fixed trough fasteners. Some care in crown fixing is necessary to ensure that at sheet overlaps the two sheets fit closely and also that the fasteners are not driven home so tightly as to deform the profile of the sheet.

Secondary fasteners, which have a shorter tail than primary fasteners, are used for fixing sheet to sheet and flashing to sheet.

Wall cladding, walling and wall sheeting Profiled steel wall sheeting is usually fixed with the profile vertical for the convenience in fixing to horizontal sheeting rails fixed across columns. The sheeting may be fixed horizontally for appearance. Horizontal fixed sheeting will be just as effective as a weather-resistant cladding as vertically fixed sheet because the overlap of the sheets and the slope of the profiles will shed water running down the face of the walling. There is usually some additional steel support required to provide a fixing for horizontally fixed wall sheeting to give support particularly between widely spaced columns.

Because wall sheeting or cladding does not have to support the weight of snow or loads from those fixing or repairing roofs, it may have a less deep profile than roof sheets of similar unsupported span. A typical trapezoidal wall sheeting profile is illustrated in Fig. 58. The pitch of profiles, i.e. the distance between the centre lines of profiles, does not have to be the same for wall sheeting as that for roofing.

Mineral wool insulation, in the form of mat or quilt, for wall sheeting will be given support by spacer bars fixed between the wall sheeting and inner lining panels at sufficiently close centres to ensure that the insulant does not sag and make a break in the insulation.

Wall cladding is usually of the same system as that for roof cladding.

Either for appearance or to provide a robust type of lower walling to withstand knocks, a solid wall upstand to the lower part of walls may be constructed as illustrated in Fig. 65. The cavity wall is raised on reinforced concrete ground beams up to the sill level of side wall windows.

Gutters For appearance and as a protection against corrosion, cold formed organically coated steel eaves gutters are used. The gutters which may be laid level or more usually at a slight fall to rainwater pipes are supported by steel brackets screwed to eaves purlins under steel eaves closers, fixed under roof cladding and over wall cladding as illustrated in Figure 59. A separate steel eaves closer is fixed under roof sheeting as shown in Fig. 61. Valley gutters and parapet wall gutters may be of galvanised steep strip, cold formed to shape, with the inside of the gutter painted with bitumen as protection against corrosion. These gutters are supported on steel brackets fixed over purlins. Valley and parapet wall gutters should have an under-lining of insulation which may be half the thickness of insulation for roof cladding.

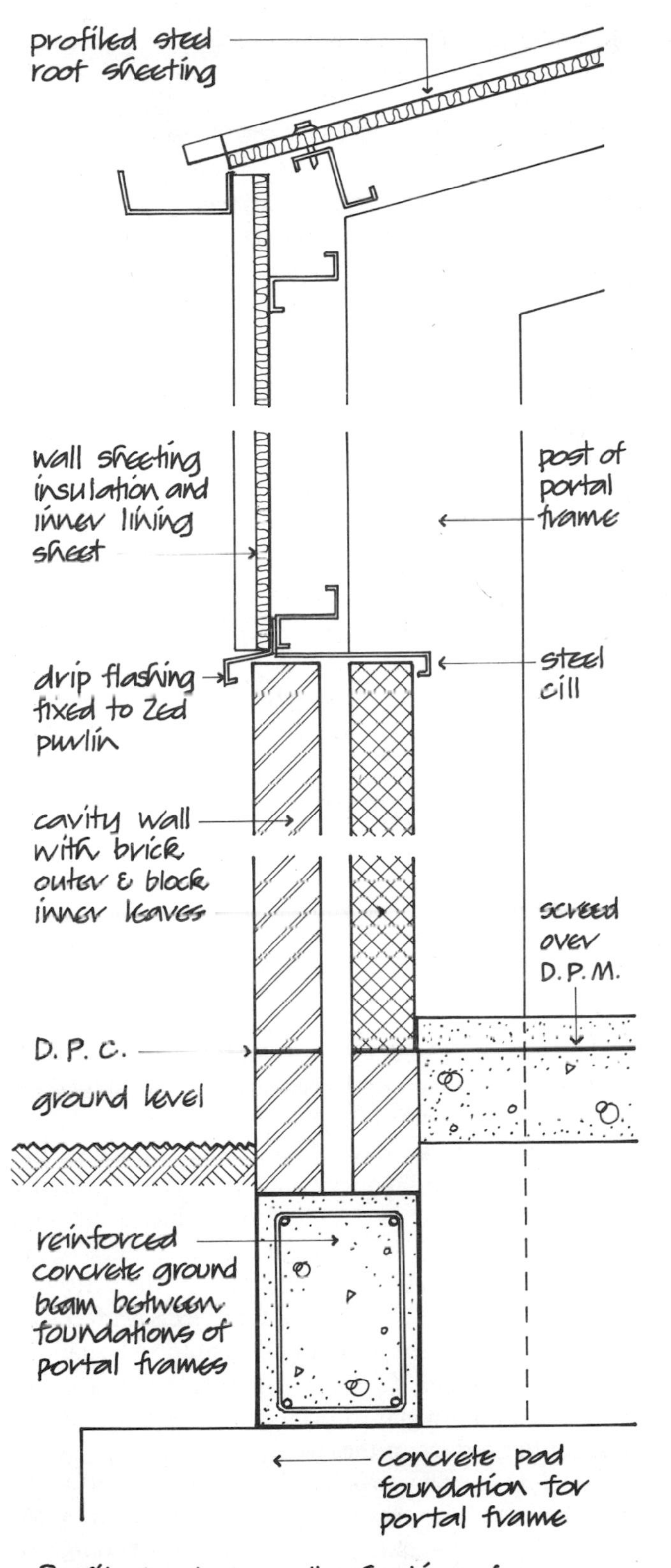

Profiled steel wall sheeting for portal frame building

Fig. 65

Ridges Ridges are covered with cold formed steel strip organically coated. The ridge may be formed to match the profile of the roof sheeting or be flat with a shaped filler to seal the spaces between the underside of the ridge and the profile of the sheeting as illustrated in Fig. 61. Wall cladding is commonly finished with a preformed steel strip closer and drip above finished external surfaces over a low concrete or brick curb upstand for appearance.

Curved sheets For appearance curved, profiled, organically coated steel sheets are used at eaves, ridges and external angles of wall cladding. The profile of the curved sheets must match the profile of the sheeting to roof and wall cladding that it is to fit to at eaves. The usual internal radius of curve is from 400 to 1200. Either plain curved sheets or laterally embossed sheets are produced. The shallow embossing of the sheets with regularly spaced sinkings and protrusions between profiles tends to assist and regularise the curving process. To an extent the shallow embossing detracts from the smooth transition between adjacent surfaces that the curvature is designed to emphasise. Figure 66 is an illustration of curved sheets to eaves. Where curved eaves sheets are used it is often practice to form a secret gutter.

Under purlin lining system To provide a flush soffit to roof cladding for appearance, it is usual to fix the inner lining and insulation under the purlins between roof

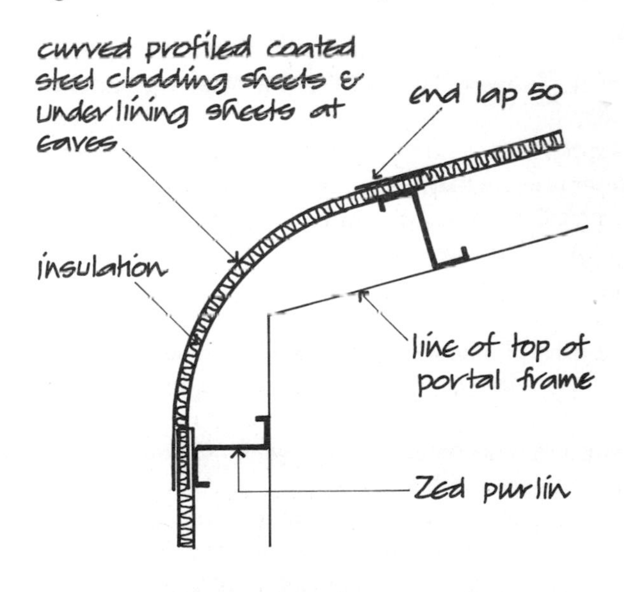

Curved profiled steel cladding

Fig. 66

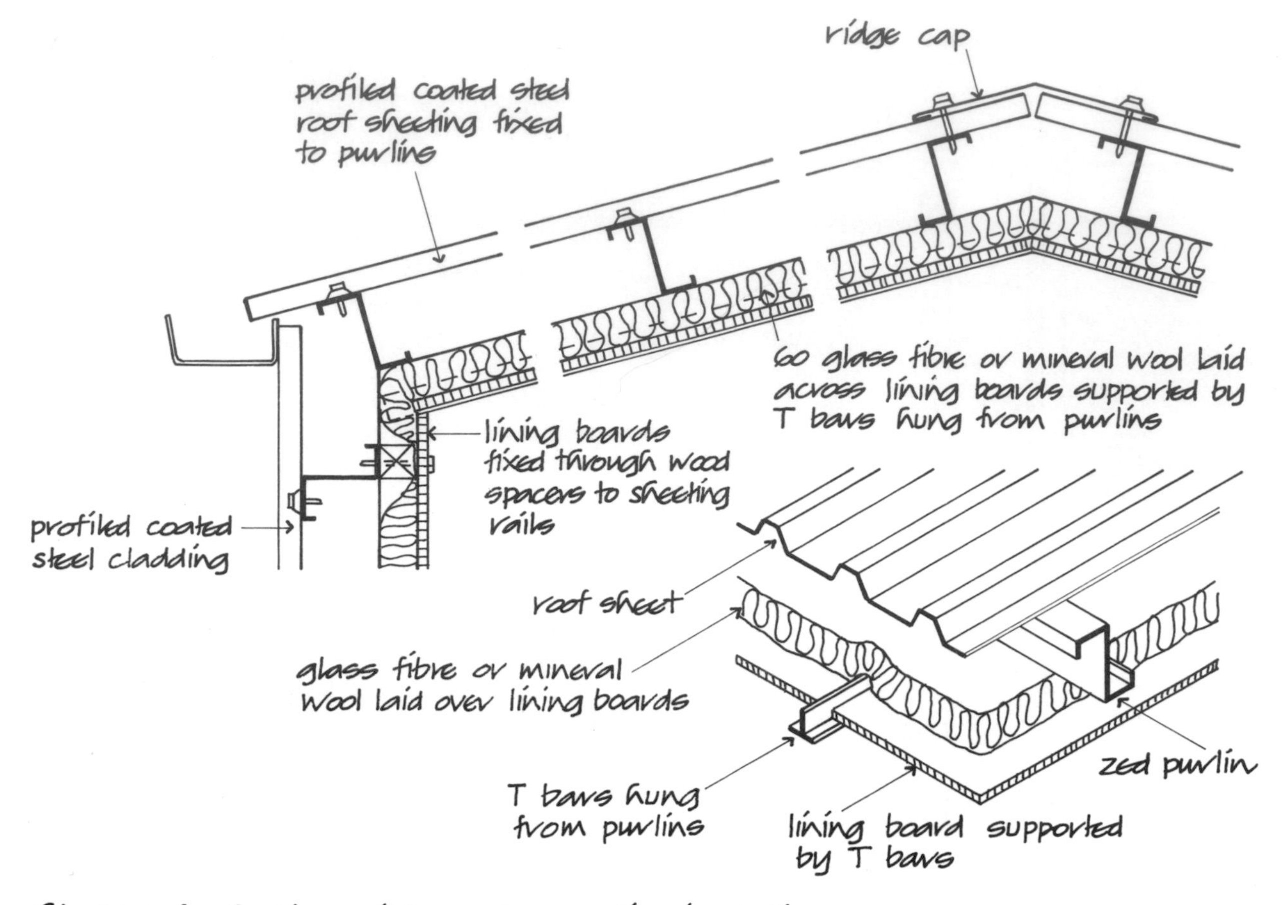

Fig. 67

frames. The roof sheeting is screwed directly to the top of the purlins. The soffit lining of preformed boards, such as plasterboard, is supported by metal T-section supports at centres to suit the boards. The T-section supports are fixed to brackets screwed to the underside of purlins. Mineral wool insulation is laid over the lining boards as illustrated in Fig. 67.

Profiled aluminium roof and wall cladding In common with other metals aluminium, on exposure to atmosphere, corrodes to form a thin coating of oxide on its surface. This oxide coating, which is integral with the aluminium, adheres strongly and being insoluble, protects the metal below from further corrosion so that the useful life of aluminium is 40 years or more.

Aluminium is a lightweight, malleable metal with poor mechanical strength, which can be cold formed without damage. Aluminium alloy strip is cold rolled as corrugated and trapezoidal profile sheets for roof and wall cladding. The sheets are supplied as metal mill finish, metal stucco embossed finish, pre-painted or organically coated.

Mill finish is the natural untreated surface of the metal from the rolling mill. It has a smooth, highly reflective metallic silver grey finish which dulls and darkens with time. Variations in the flat surfaces of the mill finish sheet will be emphasised by the reflective surface. A stucco embossed finish to sheets is produced by embossing the sheets with rollers to form a shallow, irregular raised patterned finish that reduces direct reflection and sun glare and so masks variations in the level of the surface of the sheets.

A painted finish is provided by coating the surface of the sheet with a passivity primer and a semi-gloss acrylic or alkyd-amino coating in a wide range of colours.

A two coat PVF acrylic finish to the sheet is applied by roller to produce a low-gloss coating in a wide range of colours.

Figure 68 is an illustration of profiled aluminium roof and wall sheeting, fixed over rigid insulation boards bonded to steel lining trays, to a portal steel frame.

Aluminium sheeting, which is more expensive as roof and wall cladding, is used for its greater durability particularly where humid internal atmospheres might cause early deterioration of coated steel sheeting.

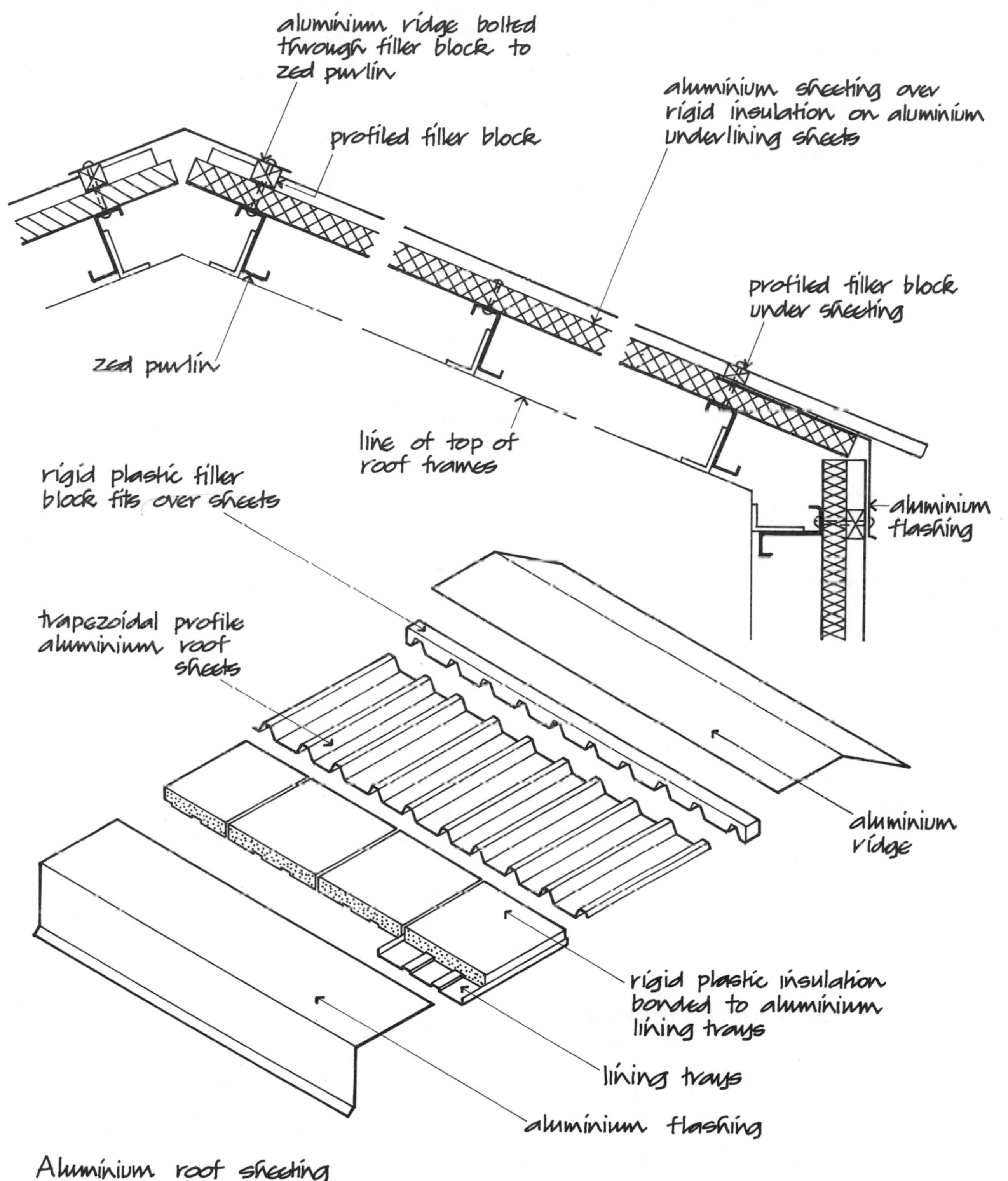

Fig. 68

FIBRE CEMENT PROFILED CLADDING

Asbestos cement sheets

The original fibre cement corrugated sheets were manufactured by felting asbestos fibres with alternate layers of cement and water. The wet mix of asbestos and cement was pressed into corrugated sheets which were steam cured to accelerate hardening of cement. The earliest corrugated asbestos cement sheets, which were first imported into this country in 1910, were manufactured here soon afterwards. For the following 50 years these sheets were extensively used for roof and side wall cladding for a variety of single-storey buildings used in agriculture, manufacturing and for storage.

The advantage of asbestos cement is that it is, unlike steel, non-corrosive, unaffected by atmospheric pollution, maintenance free and has a useful life of 40 years or more. The natural drab, light grey cement colour of the sheets does not make an attractive finish to buildings. Corrugated sheets tend to weather with irregular dirt staining on the sides of the corrugations and algae and lichen growth may flourish on the surface of the sheets in persistently damp conditions.

Because of the necessary thickness of the material it is not possible to make a close fitting at end laps, which will allow wind penetration and the fittings at ridge and eaves look somewhat lumpy and ugly as compared to the neat finish possible with thinner metal strip. Nonetheless corrugated asbestos cement sheets continued to be used for many years for the considerable advantage of low initial cost, durability and freedom from maintenance.

For appearance a range of coloured asbestos cement sheets is produced through the application of acrylic coating.

Some years ago it was established that the continuing inhalation of airborne asbestos fibres could, over time, become a serious health hazard particularly to those employed in the manufacture and use of asbestos fibre-based materials. The facts are that blue asbestos fibre, principally from South Africa, is a much more serious hazard than white asbestos fibre. A much publicised, hysterical outcry has since condemned all asbestos fibre for any use whatsoever and in all conditions regardless of good sense and statistical evidence.

In consequence asbestos fibre reinforced cement sheets are no longer used in this country. As a substitute, fibre reinforced cement sheets are produced in a range of corrugated and profiled sheets both for replacement for damaged asbestos cement sheets and also for new work. These fibre cement sheets are manufactured from cellulose and polymeric fibres, Portland cement and are water pressed to profiles the same or very similar to the original asbestos cement sheets. A range of acrylic coated coloured sheets is produced.

Corrugated fibre cement sheets

These sheets are pressed to a sinusoidal or corrugated profile with a 73 or a 146 pitch of corrugations which coincides with the original 3 inch and 6 inch imperial measure. These two corrugated profiles are still fairly extensively used for replacement and repair work, and

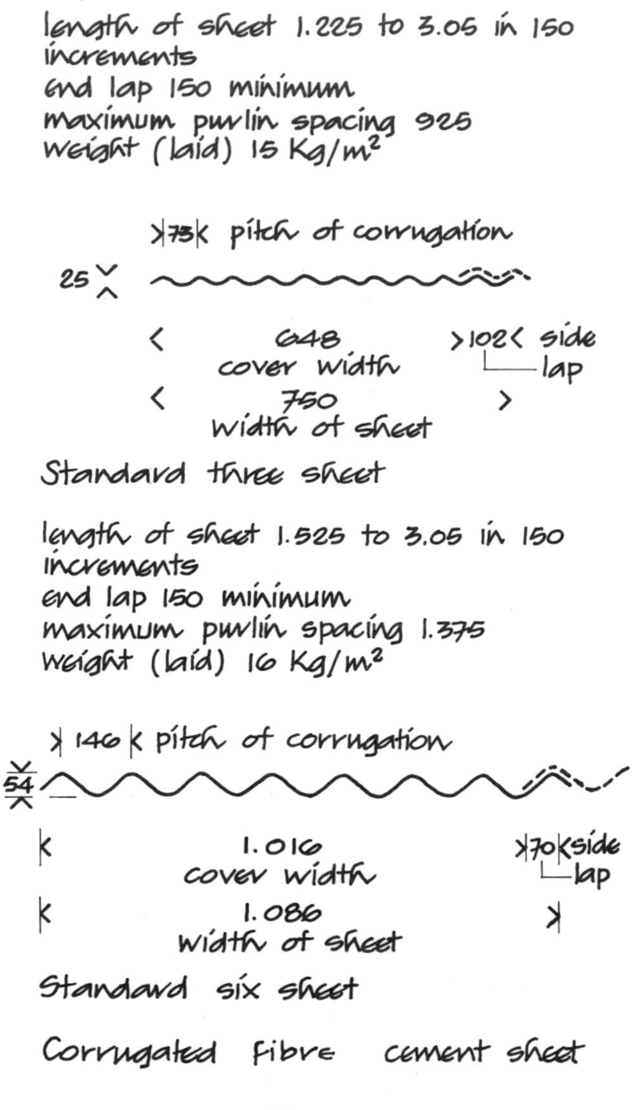

length of sheet 1.225 to 3.05 in 150 increments
end lap 150 minimum
maximum purlin spacing 925
weight (laid) 15 Kg/m²

73 pitch of corrugation
25
648
cover width
102 side lap
750
width of sheet

Standard three sheet

length of sheet 1.525 to 3.05 in 150 increments
end lap 150 minimum
maximum purlin spacing 1.375
weight (laid) 16 Kg/m²

146 pitch of corrugation
54
1.016
cover width
70 side lap
1.086
width of sheet

Standard six sheet

Corrugated fibre cement sheet

Fig. 69

for new cladding to single-storey buildings for the advantage of freedom from maintenance and freedom from corrosion both from moisture externally and condensation moisture from warm inside air. These sheets are heavier and require closer centres of support from purlins and rails than the deeper profile steel sheets.

Figure 69 illustrates the principal dimensions of profile 3 and profile 6 corrugated fibre cement sheets.

A range of fittings is made for finishes at ridge and eaves as illustrated in Fig. 70. The ridge is covered with a two piece fitting, each half of which fits over the corrugations of the two slopes. Eaves closer and eaves filler pieces fit under the corrugations of the eaves sheets.

Corrugated fibre cement sheet is usually at a pitch or slope of at least 10° to the horizontal but can be fixed at a pitch of not less than 5° in sheltered positions. Figure 71 is an illustration of corrugated fibre cement sheets fixed as roof and side walling to a single-storey, single-bay framed building which does not require heating. The sheeting is fixed with galvanised hook bolts to angle purlins with end lap between sheets of 150. The two piece ridge is hook bolted through the sheeting and the eaves filler is bolted to the ends of the sheeting.

Where the interior of the building is to be heated a system of over purlin insulation with under lining sheets, similar to that used for profiled steel sheeting, is used.

A variety of fixings is available to suit the various purlins used to support the sheets as illustrated in Fig. 72.

Figure 73 is an illustration of the use of profiled fibre cement sheets with mineral wool quilt insulation laid over steel underlining sheets with galvanised steel spacers on cold formed purlins.

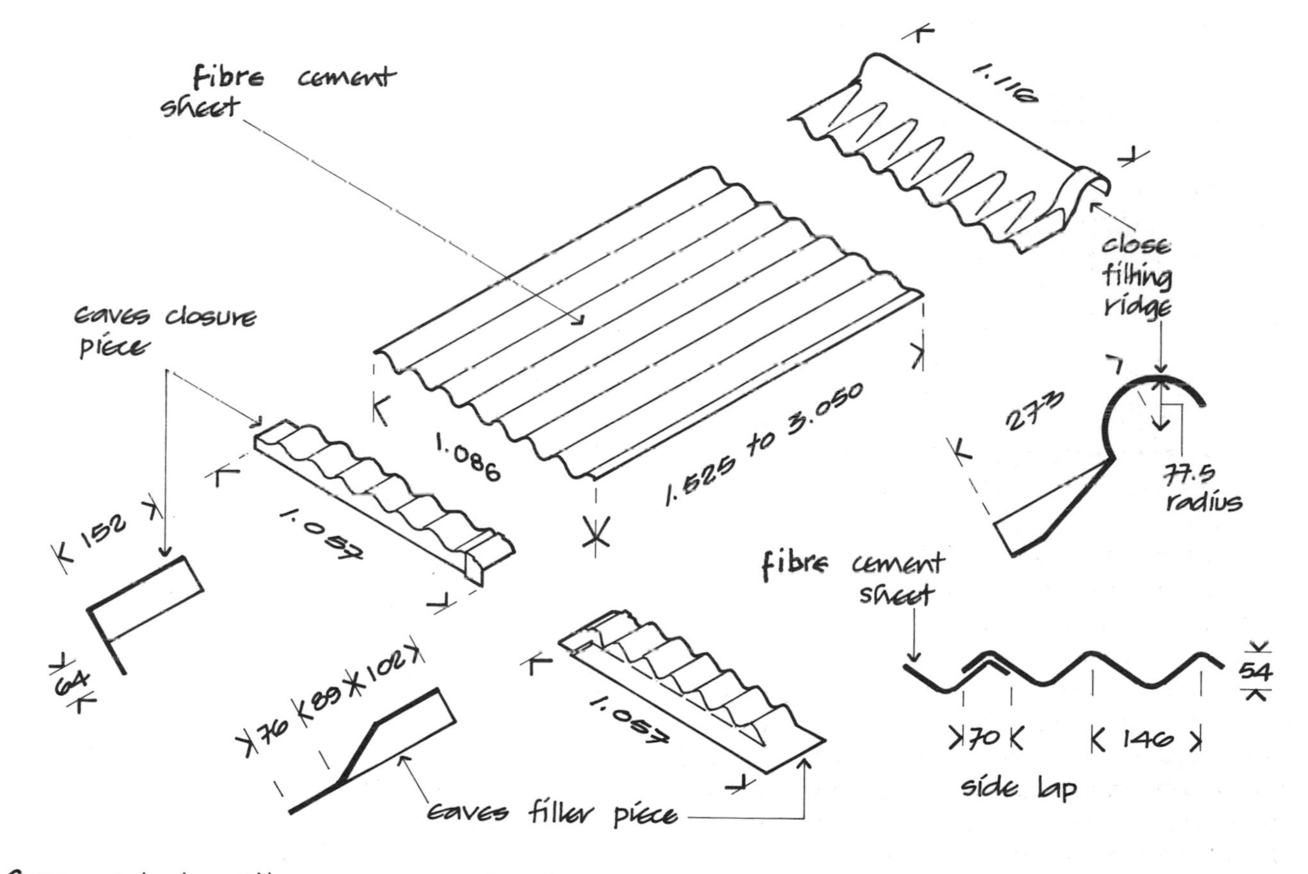

Fig. 70

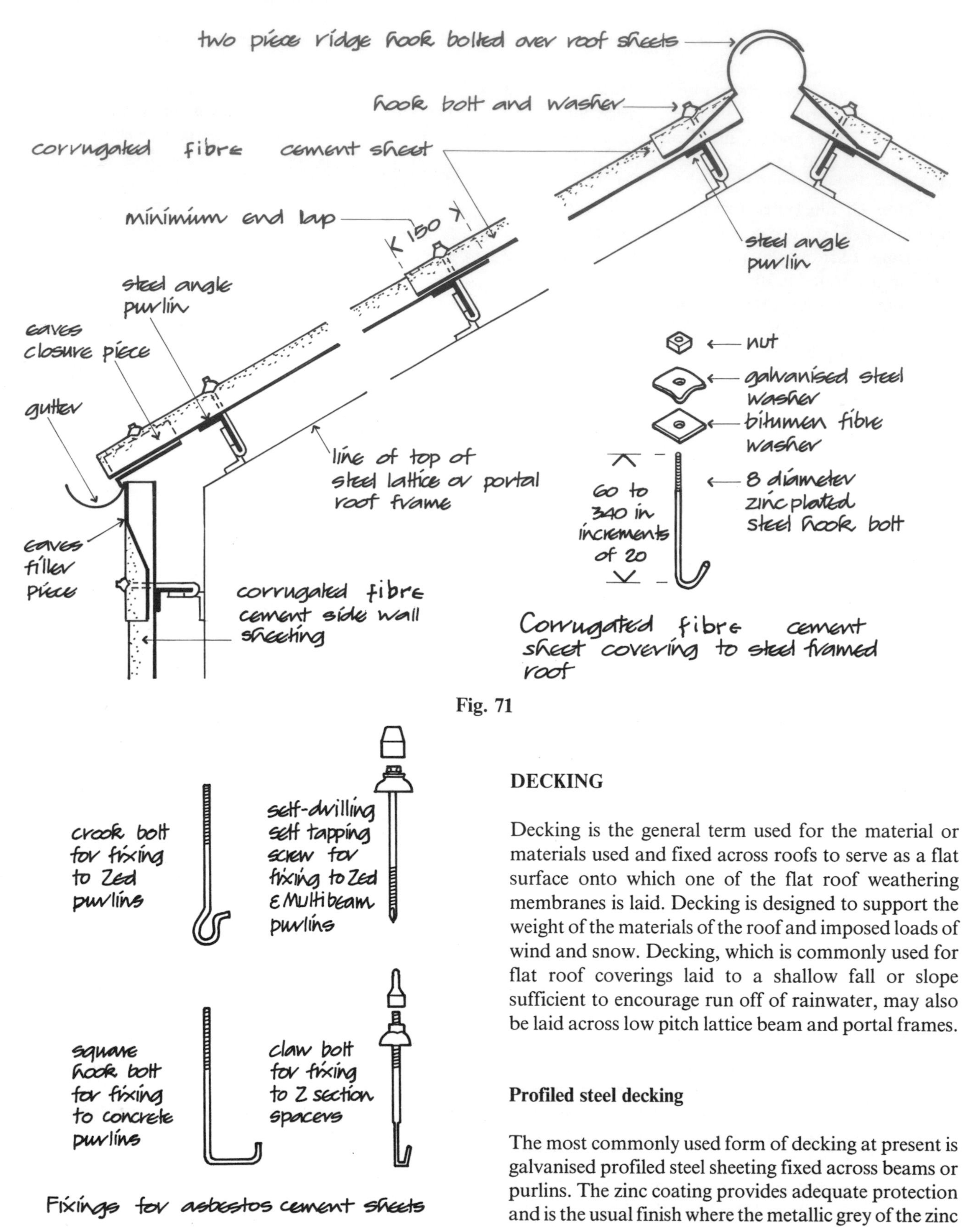

Fig. 71

Fig. 72

DECKING

Decking is the general term used for the material or materials used and fixed across roofs to serve as a flat surface onto which one of the flat roof weathering membranes is laid. Decking is designed to support the weight of the materials of the roof and imposed loads of wind and snow. Decking, which is commonly used for flat roof coverings laid to a shallow fall or slope sufficient to encourage run off of rainwater, may also be laid across low pitch lattice beam and portal frames.

Profiled steel decking

The most commonly used form of decking at present is galvanised profiled steel sheeting fixed across beams or purlins. The zinc coating provides adequate protection and is the usual finish where the metallic grey of the zinc coat is an acceptable finish to the exposed underside of

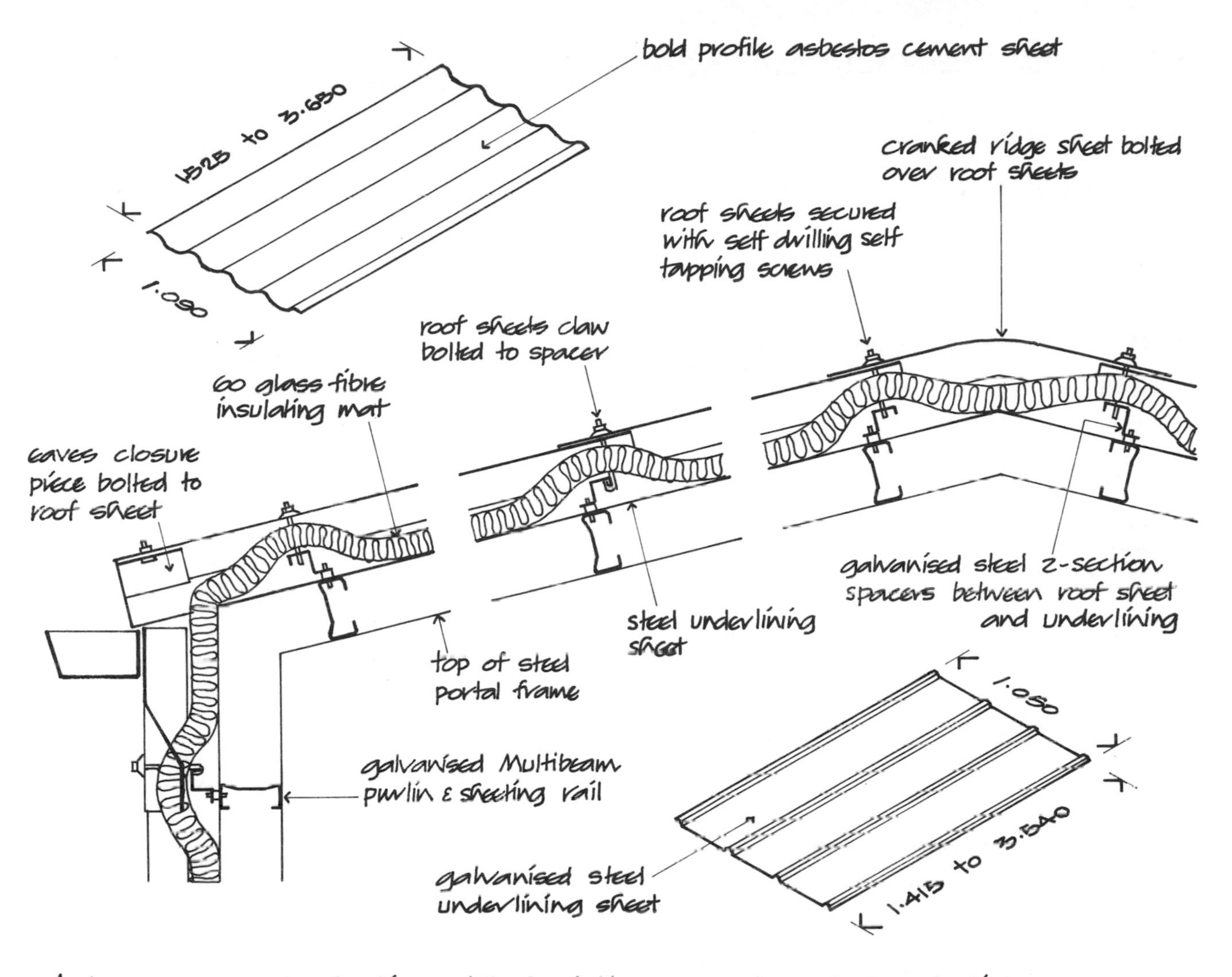

Fig. 73

the decking. As an alternative, the underside of the decking may be primed ready for painting or be coated with a paint finish.

Trapezoidal profile steel decking with depths of up to 200 are produced as decking for spans of up to 12.0 between structural frames or beams. As stiffening against buckling, ribs are pressed in the ridges, troughs and sides of the profiles as illustrated in Fig. 74. The decking is secured with screws driven through cold formed purlins or by self-tapping screws to holes in beams and roof frames.

The steel decking provides support for rigid insulation boards which are either bedded in bitumen on the ridges of the decking or secured with self-drilling screws, depending on the anticipated wind uplift. The weathering membrane is then bonded to the insulation boards as illustrated in Fig. 75. This is a form of warm roof or warm construction as the roof frame, intermediate supports and the steel decking are below the insulation and, as such, will be maintained at inside air temperature. Where there is likely to be appreciable moisture vapour pressure drive from warm inside air towards colder outside air, it is practice to bond a layer of some material such as bitumen felt or polythene sheet across the steel decking below the insulation to act as a vapour check.

Profiled steel decking is used as the support for insulation and a weather finish of built-up bitumen felt,

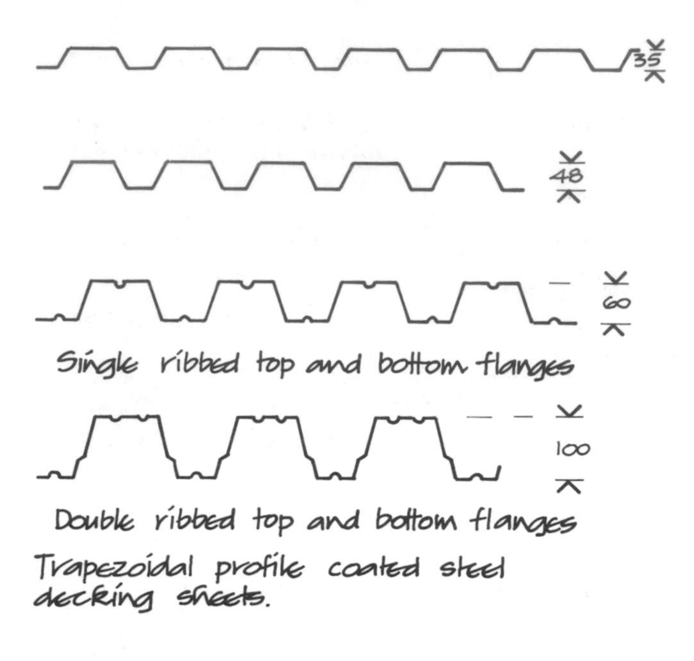

Fig. 74

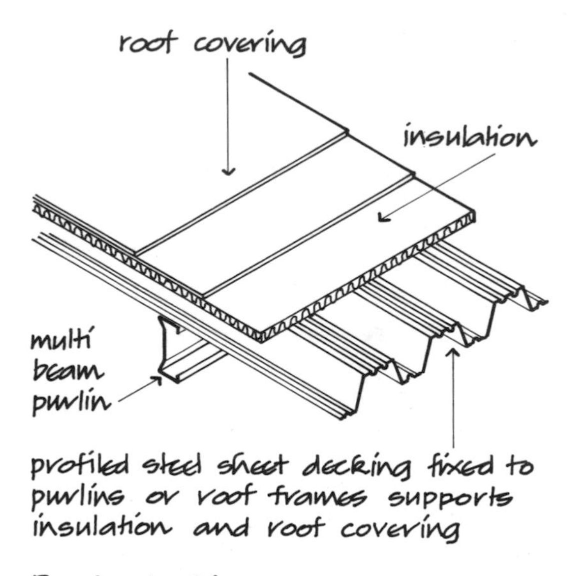

Fig. 75

high performance felt or a single ply membrane for flat or low pitch roofs for the advantage of widely spaced supports that is possible with deep profile decking.

The trapezoidal profile steel decking illustrated in Fig. 76 is fixed across the secondary beams of a steel framed flat roof structure with self-tapping fasteners. A felt vapour check is bonded to the decking with bitumen. Rigid insulation boards are bonded to the vapour check either with bitumen or self-drilling self-tapping screws and flat washers as necessary to resist wind suction uplift. The roof is finished with built-up bitumen felt as a weathering surface.

The requisite fall or slope for rainwater run off is provided by taper section insulation boards.

Composite steel decking and cladding

The system of steel decking and mineral fibre insulation illustrated in Fig. 77 combines the long span advantage of steel decking, a high degree of insulation and profiled steel cladding for low pitch roofs. The profiled steel decking has deep web stiffening longitudinally for spans of up to 12.0, for fixing between widely spaced beams or roof frames without intermediate support. Lattic steel spacer bars are fixed to the decking over a polythene vapour check to maintain the thickness of a thick layer of fibre insulation. The profiled PVC or acrylic coated steel cladding sheets are supported by the spacer bars, under the longitudinal ridges of the cladding, which has shallow transverse stiffening ribs to strengthen the wide troughs of the profile.

This roofing system is designed specifically for the high insulation that is possible due to the spacer bars and the life expectancy to first maintenance of the PVC or acrylic coating. Fibre insulation thicknesses of 160, 220 and 270 are available with this system.

Wood wool slab decking

Wood wool slabs are made from seasoned wood fibres bound together under pressure with Portland cement. The finished slabs which have coarse, open textured surfaces, moderate compressive strength and thermal resistance and good resistance to rot and fungal growth, are used as roof deck for flat roofs to support the roof covering and provide some degree of insulation. Wood wool slabs are made in thicknesses of 50, 75, 100 and 125, widths of 600 either as plain slabs or channel interlocking slabs. The material of the slabs is combustible but not readily ignited and the spread of flames is low. Plain slabs are used as a deck to timber-framed roofs with the slabs supported at up to 600 centres. Channel interlocking slabs are manufactured with steel channels in the long edges of the slabs as reinforcement for spans of up to 4.0 for normal roof

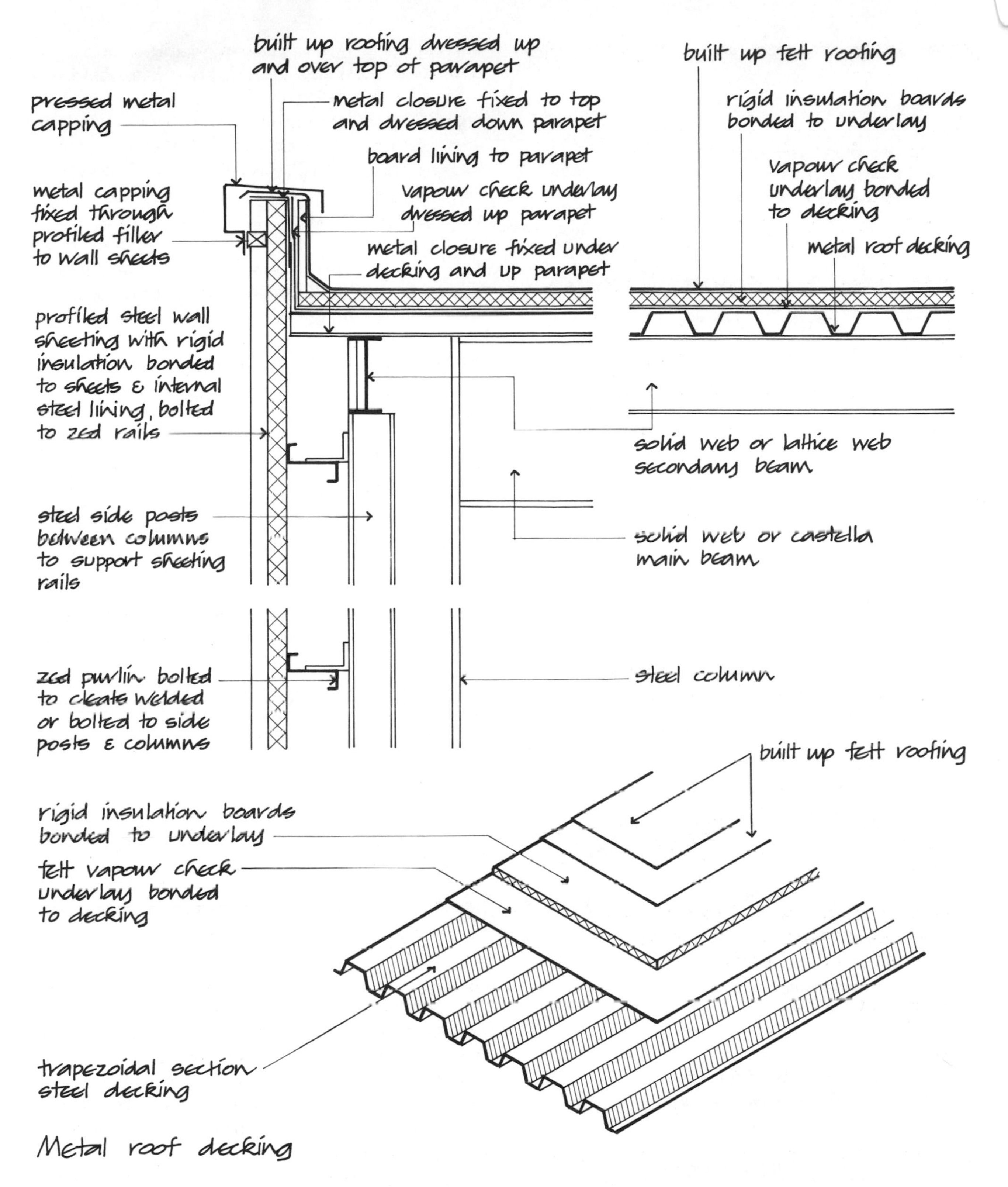
built up roofing dressed up and over top of parapet
pressed metal capping
metal closure fixed to top and dressed down parapet
board lining to parapet
metal capping fixed through profiled filler to wall sheets
vapour check underlay dressed up parapet
metal closure fixed under decking and up parapet
built up felt roofing
rigid insulation boards bonded to underlay
vapour check underlay bonded to decking
metal roof decking
profiled steel wall sheeting with rigid insulation bonded to sheets & internal steel lining, bolted to zed rails
solid web or lattice web secondary beam
steel side posts between columns to support sheeting rails
solid web or castella main beam
zed purlin bolted to cleats welded or bolted to side posts & columns
steel column
built up felt roofing
rigid insulation boards bonded to underlay
felt vapour check underlay bonded to decking
trapezoidal section steel decking
Metal roof decking

Fig. 76

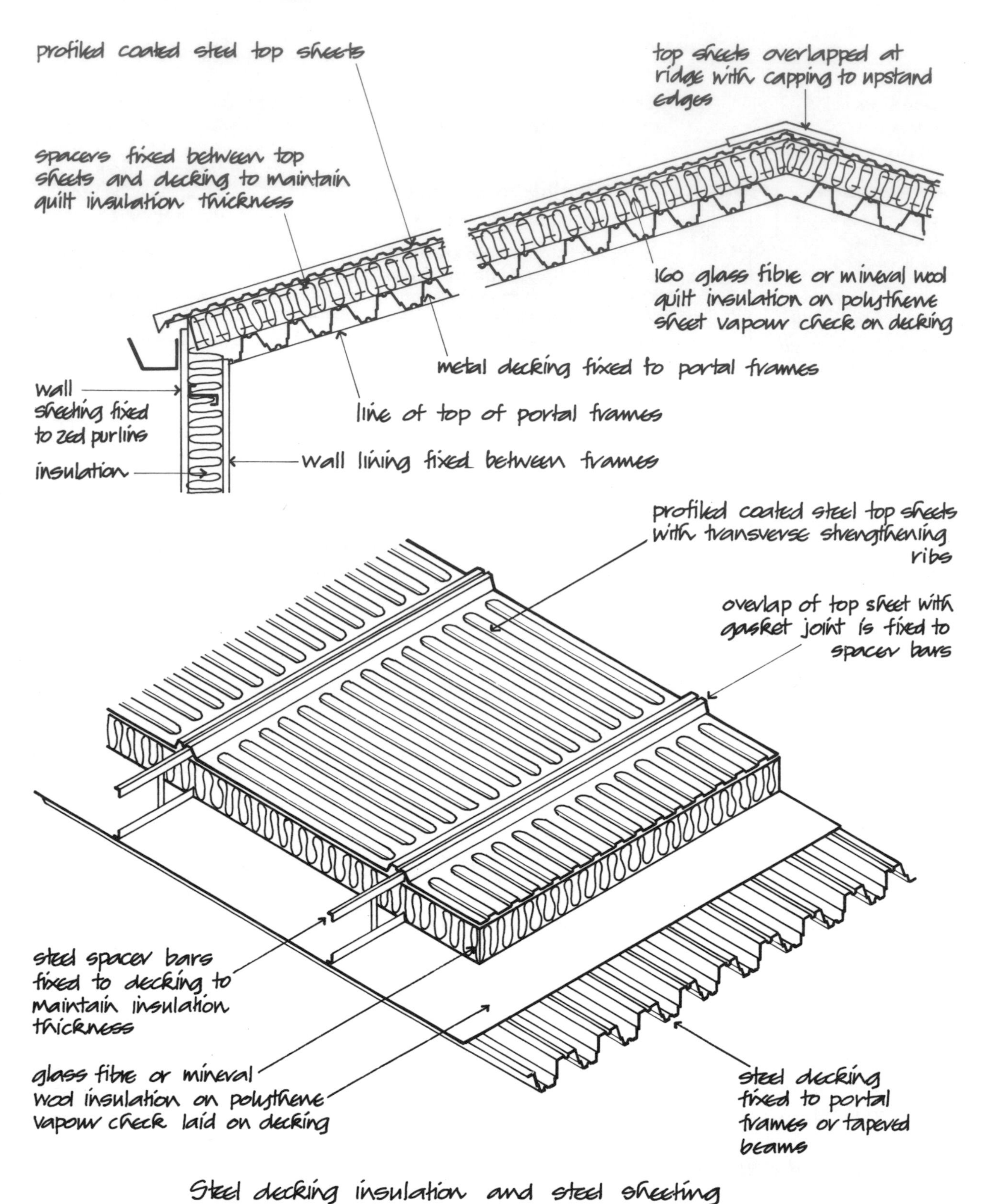
profiled coated steel top sheets
top sheets overlapped at ridge with capping to upstand edges
spacers fixed between top sheets and decking to maintain quilt insulation thickness
160 glass fibre or mineral wool quilt insulation on polythene sheet vapour check on decking
metal decking fixed to portal frames
wall sheeting fixed to zed purlins
line of top of portal frames
insulation
wall lining fixed between frames
profiled coated steel top sheets with transverse strengthening ribs
overlap of top sheet with gasket joint is fixed to spacer bars
steel spacer bars fixed to decking to maintain insulation thickness
glass fibre or mineral wool insulation on polythene vapour check laid on decking
steel decking fixed to portal frames or tapered beams
Steel decking insulation and steel sheeting

Fig. 77

loads. Another system of wood wool slabs employs either 'T' plates or I-section supports as reinforcement. Tee section cold formed steel supports are fixed with clips and plates that are screwed, welded or shot fired into purlins or secondary beams and the wood wool slabs are laid between the T-sections as illustrated in Fig. 78. The joints between the slabs are covered with hessian scrim in a slurry of cement and sand. Insulation boards can be bonded or screwed to the slabs to provide additional insulation and as a base for the roof covering.

The I-section cold formed steel supports for slabs fit to grooves cut in the long edges of the slabs. The I-section supports are fixed to each supporting beam by screwing, welding or shot firing. The joints between the slabs are covered with hessian scrim in a sand and cement slurry as shown in Fig. 79.

A disadvantage of the steel channel reinforcement and the cold formed T- and I-section supports is that they will act as cold bridges to the transfer of heat and may encourage condensation and dirt staining on the soffit of plastered slabs.

To provide a fall for roof drainage the slabs can be laid across taper or sloping beams, a screed to falls can be spread over the surface of the slabs, or taper section insulation boards can be laid across the slabs.

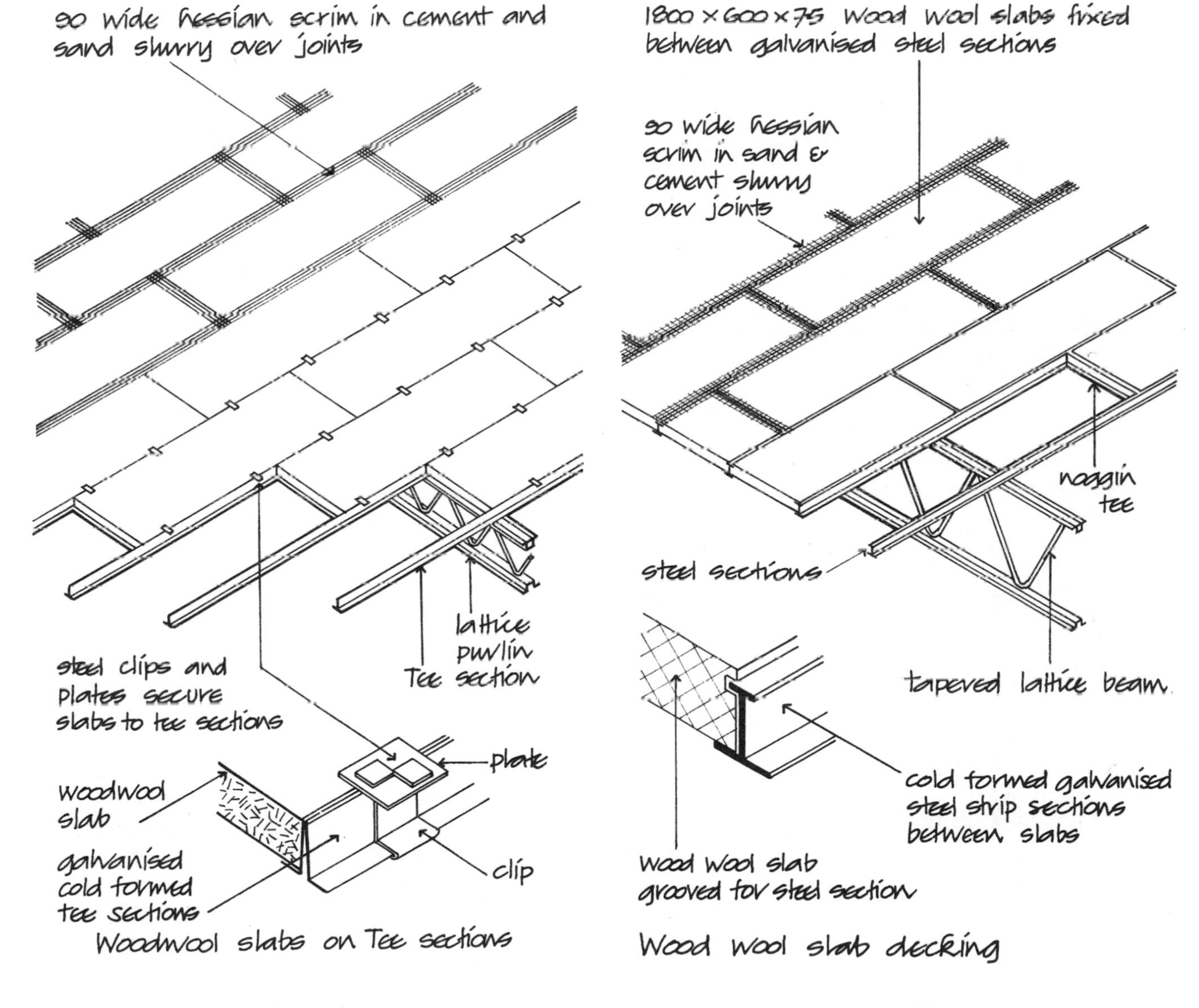

Fig. 78

Fig. 79

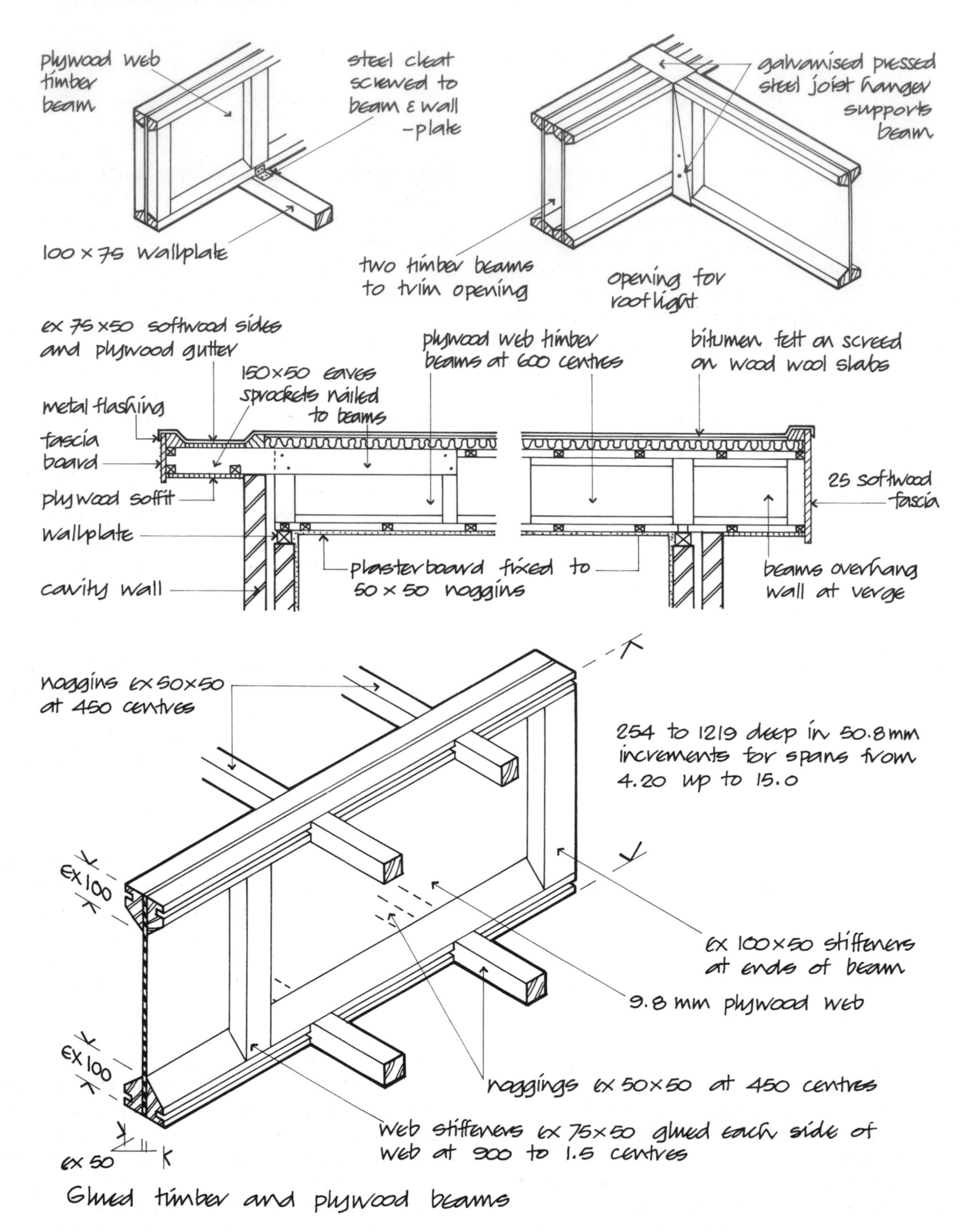
plywood web timber beam
steel cleat screwed to beam & wall -plate
100 × 75 wallplate
galvanised pressed steel joist hanger supports beam
two timber beams to trim opening
opening for rooflight
ex 75 × 50 softwood sides and plywood gutter
150 × 50 eaves sprockets nailed to beams
metal flashing
fascia board
plywood soffit
wallplate
cavity wall
plywood web timber beams at 600 centres
bitumen felt on screed on wood wool slabs
25 softwood fascia
plasterboard fixed to 50 × 50 noggins
beams overhang wall at verge
noggins ex 50 × 50 at 450 centres
254 to 1219 deep in 50.8 mm increments for spans from 4.20 up to 15.0
ex 100
ex 100
ex 100 × 50 stiffeners at ends of beam
9.8 mm plywood web
noggings ex 50 × 50 at 450 centres
web stiffeners ex 75 × 50 glued each side of web at 900 to 1.5 centres
ex 50
Glued timber and plywood beams

Fig. 80

Wood wool slabs are less used than they were since the advent of profiled steel as decking.

Timber decking

The advantage of the use of timber beams as the structural deck for flat roofs is that the timber requires no maintenance and does not suffer corrosion in humid atmospheres, e.g. in swimming pools. Because the maximum span of timber beams is about 15.0 and the beams are more expensive than comparable standard lattice steel beams, the use of timber beams is generally limited to medium span roofs where conditions such as humid atmospheres justify their use.

The timber beam deck illustrated in Fig. 80 consists of timber beams fabricated with plywood webs and softwood section top and bottom booms, glued together with timber web stiffeners and noggin pieces. The beams are fixed at centres to suit wood wool slabs that serve as surface to the deck and insulation. To provide additional insulation, a layer of taper insulation board is bonded to the slabs to provide drainage falls. The beams may be exposed or covered with a soffit of plasterboard.

Lightweight concrete slab decking

Lightweight reinforced concrete slabs are made from a mix of Portland cement, finely ground sand and lime. The materials are mixed with water and a trace of aluminium powder is added. The mix is cast in moulds around reinforcing bars and auto claved to cure and harden the slabs. The effect of the trace of aluminium powder is to entrain minute bubbles of gas in the concrete which is lightweight and has better resistance to the transfer of heat than dense concrete. The lightweight slabs are designed specifically for use as deck material for roofs where the resistance to damage by fire is an advantage.

The slabs are laid across steel or concrete flat or low pitch structures as a deck as illustrated in Fig. 81. The slabs are secured with steel straps cast into or fixed to the structure and reinforced with continuity steel bars bedded to the joints. The slabs can be used by themselves as a deck for flat roof covering or with a layer of insulation under the covering.

FLAT ROOF WEATHERING

With the now common use of profiled sheeting to roofs, with a slope or pitch of as little as 2½° to the horizontal, the terms 'flat roof' and 'pitched roof' are no longer clearly defined. A flat roof has been used to describe a roof with a finished weathering surface of up to 5° to the horizontal and a pitched roof as a roof with a slope of over 5° to the horizontal. In current roof terminology a pitched roof has a weathering surface of more than 10° to the horizontal, a low pitched roof from 5° to 10° to the horizontal and a very low pitch or flat roof a slope of up to 5° to the horizontal.

There is no economic or practical advantage in the use of a flat roof structure unless the roof is used as a deck for leisure, recreation or storage purposes. A flat roof structure is less efficient structurally than a pitched roof structure and there is often little, if any, saving in unused roof space as compared to a low pitch frame structure. The inevitable deflection of horizontal beams that would cause ponding of rainwater on flat roofs requires the construction of falls on the roof to clear rainwater to outlets. The many failures of flat roofs that have occurred over the years have not tended to recommend the use of flat roofs.

Nonetheless it has been fashionable for many years to construct large buildings with flat roofs for the sake of the horizontal line that was in favour and to avoid too large an expanse of visible roof.

Recent improvements in flat roof weathering membranes and careful detailing have made flat roof coverings a viable alternative to sheet metal profiled sheets.

It is generally accepted practice for flat roofs to be constructed with a finished weathering surface that has a fall or slope of at least 1 in 80 to rainwater outlets so that rainwater will run off towards outlets. This fall is often increased to 1 in 40 to make allowance for deflection of beams under load and inaccuracies in construction.

Due to the self-weight of a roof structure and the imposed loads of wind and snow, every horizontal structure will deflect or sag at mid-span between points of support. It would, therefore, seem logical to accept this inevitable deflection as a means of providing the shallow fall or slope necessary to drain rainwater to outlets at mid-span. This sensible approach to design in the utilisation of an inevitable structural effect, to drain a roof surface, is not generally accepted because it is usually inconvenient to have rainwater down pipes

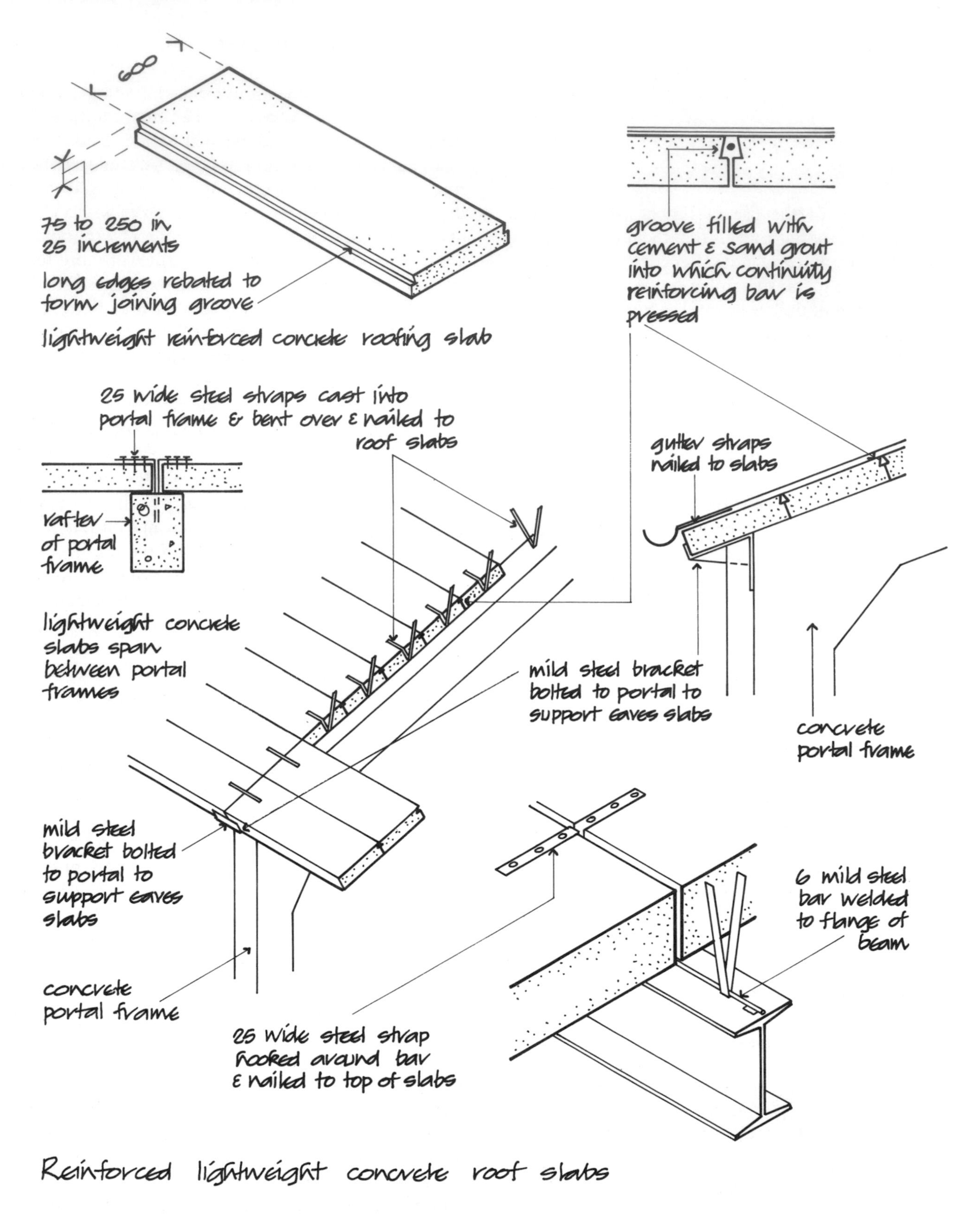
600
75 to 250 in
25 increments
long edges rebated to
form joining groove
lightweight reinforced concrete roofing slab
groove filled with
cement & sand grout
into which continuity
reinforcing bar is
pressed
25 wide steel straps cast into
portal frame & bent over & nailed to
roof slabs
rafter
of portal
frame
gutter straps
nailed to slabs
lightweight concrete
slabs span
between portal
frames
mild steel bracket
bolted to portal to
support eaves slabs
concrete
portal frame
mild steel
bracket bolted
to portal to
support eaves
slabs
6 mild steel
bar welded
to flange of
beam
concrete
portal frame
25 wide steel strap
hooked around bar
& nailed to top of slabs
Reinforced lightweight concrete roof slabs

Fig. 81

running down inside a building at the middle of a clear span of structures, as the pipes will probably constrain freedom of layout of activities on floors and are unsightly. Rainwater pipes could be run with a shallow fall under roofs to down pipes fixed to columns or walls. Rainwater pipes run to shallow falls need sealed joints and are more subject to blockages than vertical pipes. The most convenient place to run rainwater down pipes is at points of support, such as columns and walls, just where the deflection of a roof is least and natural run-off of water will not occur.

Drainage and falls

For the reasons given above, horizontal, flat roof structures must have some form of construction to provide a fall or slope to the roof surface to drain water to the necessary outlets to rainwater down pipes fixed to points of support. The choice of a means of construction to provide a fall will depend on whether the roof surface is to fall to eaves or parapet gutters or to have a fall in two directions to central outlets as illustrated in Fig. 82. The junction of falls in two directions that form a shallow valley is termed a current. The more straightforward one-direction fall can be constructed with some form of tapered lattice beam with the top boom sloping at 1 in 40 or with tapered firring pieces of wood or tapered insulation boards laid over the structure to provide the necessary falls. The two direction fall is more complicated to construct because of the need to mitre the ends of tapered firring pieces or tapered boards to form the junction of falls running at right angles to each other. A wet screed can be laid and finished with cross falls without difficulty.

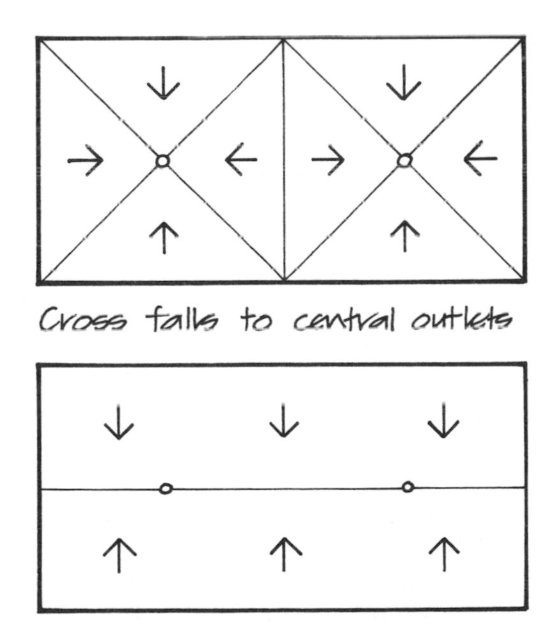

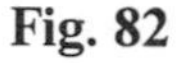
Fig. 82

Flat roof coverings are usually designed to fall directly to rainwater outlets without the use of gutters or sumps formed below the surface of the roof. The run-off of water from a flat roof is comparatively slow and regular and there is no advantage in the use of gutters or sumps to accelerate the flow of water and considerable disadvantage in the construction of gutters and sumps and the additional cost of joints necessary in the finishing of a roof covering around the edges of gutters and sumps.

A typical straight-fall rainwater outlet is illustrated in Fig. 83 where the roof falls in two directions to a level valley that is drained by internal outlets above points of support of the structure with the down pipes run against columns. The falls are provided by tapered lattice, secondary beams spanning between main beams.

The area of roof to be drained to rainwater outlets and the flow of water is determined by reference to maps of England showing areas of peak rainfall. The area of roof to be drained is usually based on the need to drain rainwater in periods of maximum rainfall. Calculations are commonly based on a fall of 75 mm per hour for two minutes which occurs with a frequency on average once every year in most of the more densely populated parts of the Midlands and South of England.

Insulating flat roofs

Warm roof The majority of flat roofs are constructed as a deck, laid across the members of the structural roof. The deck acts as support for the flat roof covering, below which a layer of some insulating material is laid. Because the insulation is above the deck, this form of roof construction is described as a 'warm roof' as the deck is insulated from the outside and will, therefore, tend to be maintained at the same temperature as the heated inside of buildings and be much less vulnerable to condensation than it would be were the insulation below the deck. The warm roof form of construction is

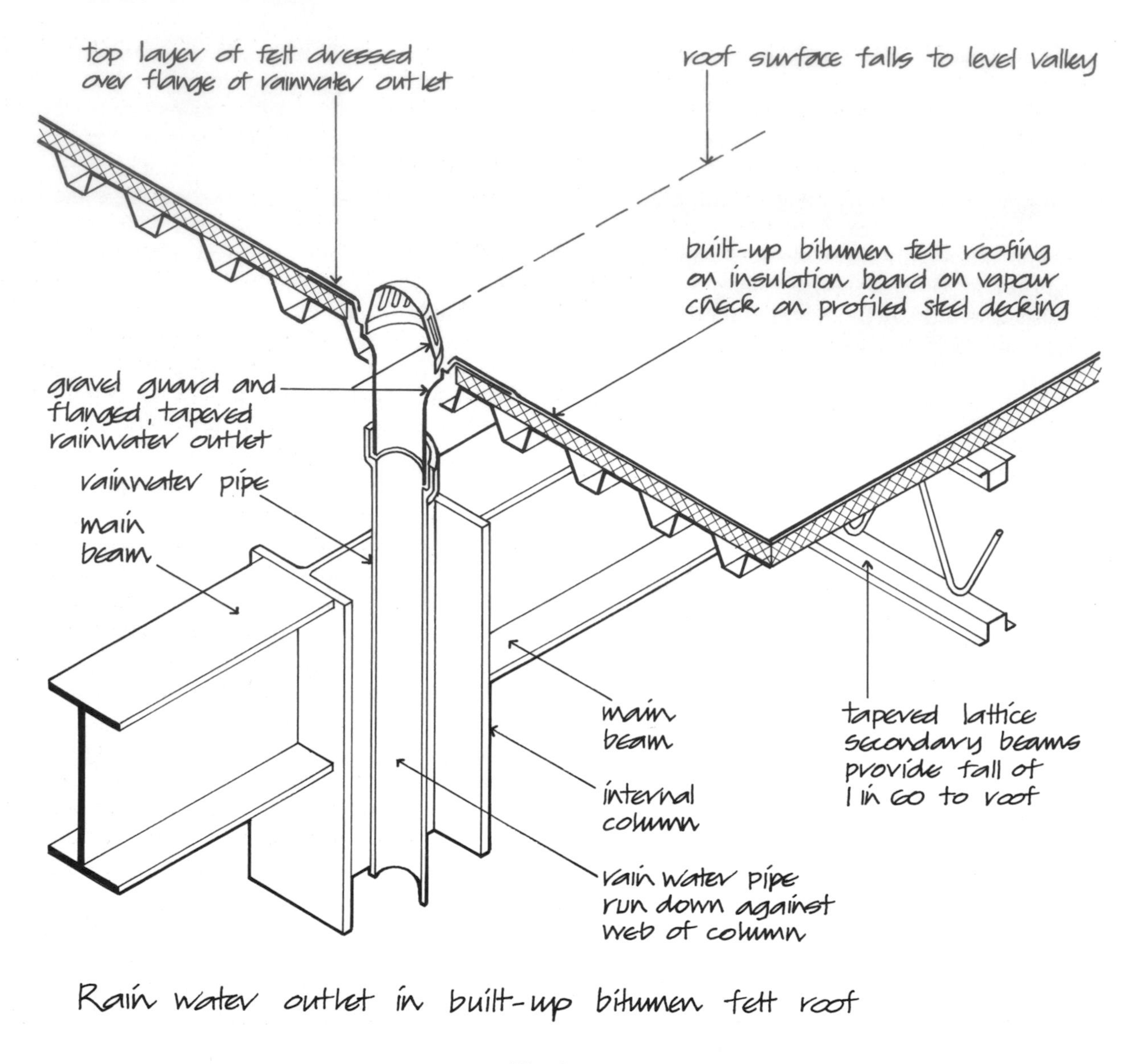

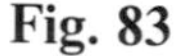
Fig. 83

of particular advantage where profiled steel decking is used as the support for the roof surface, as the insulation above the deck greatly reduces the likelihood of warm moist air from the inside condensing on the surface of the metal deck. The disadvantage of the warm roof is that the insulation under the roof covering will increase the temperature differences of the covering between day and night and so increases elongation stress and fatigue of the covering.

Ventilation Approved Document F, which gives practical guidance to meeting the requirements from Part F of Schedule 1 to the Building Regulations 1991, states that the requirements will be met if condensation in a roof and in the space above insulated ceilings is limited so that under normal conditions:

(a) The thermal performance of the insulating materials and
(b) The structural performance of the roof construction will not be substantially and permanently reduced.

The Approved Document states that it is not necessary to ventilate warm deck roofs or inverted roofs where the moisture from the building cannot permeate the insulation, and the requirements will be met by ventilation of cold deck roofs where the moisture from the building can permeate the insulation.

The effect of this guidance is that warm deck roofs may need a vapour check below the insulation layer where the vapour pressure drive from inside warm air to outside colder air may cause moisture vapour to penetrate the insulation. Cold deck roofs require cross-ventilation of the space above insulation and below the roof covering as illustrated in Fig. 55.

Insulation materials The materials most used as insulation to flat roofs are in the form of rigid boards that are laid over the structural deck of warm roofs as support for the roof covering. Mineral fibre materials in the form of wood wool and glass fibre quilts and mats are used at ceiling level laid over and supported by the ceiling finish in cold roof forms of construction.

The board materials most used are vegetable wood fibre softboards for economy and cork boards, plastic foam, polystyrene bead and extruded boards, rigid urethane (PUR) and phenolc (PIR) boards.

Materials for flat roof coverings

Mastic asphalt Asphalt is one of the traditional materials used as a waterproof surface covering for flat roofs. The original mastic asphalt was a mixture of natural rock asphalt and natural lake asphalt.

Natural rock asphalt is mined from beds of limestone which were saturated or impregnated with asphaltic bitumen thousands of years ago. The rock, which is chocolate brown in colour, is mined in France, Switzerland, Italy and Germany. The rock is hard, and because of the bitumen with which it is impregnated, it does not as readily absorb water as ordinary limestone.

Natural lake asphalt is dredged from the bed of a dried-up lake in Trinidad. It contains a high percentage of bitumen with some water and about 36%, by weight, of finely divided clay.

Asphalt is manufactured either by crushing natural rock asphalt and mixing it with natural lake asphalt or, more usually today, by crushing natural limestone and mixing it with bitumen, or a mixture of bitumen and lake asphalt, while the two materials are sufficiently hot to run together. The heated mixture of asphalt is run into moulds in which it cools and solidifies. The British Standard specification for mastic asphalt using natural rock asphalt is BS 6577 and that for limestone aggregates is BS 6925.

The solid blocks of asphalt are heated on the building site and the hot plastic material is spread and worked into position and levelled with a wood float in two coats, breaking joint, to a finished thickness of 20. The first 10 coat is spread on a separating layer of sheathing felt that is laid on the surface of the roof without a bond to the roof deck. The purpose of this layer is to isolate the asphalt from movements that will occur in the roof structure below and also to reduce blowing of asphalt by allowing lateral escape of entrapped air and moisture that would otherwise expand and cause blow holes or blisters in the surface of the asphalt. When the top surface of the asphalt has been finished level with a wood float and the asphalt is still hot, the surface is dressed with fine sand that is spread and lightly rubbed into the surface with the float. The purpose of the sand dressing is to break up the bitumen-rich top surface and so avoid unsightly crazing of the top surface that would occur as the rich bitumen surface oxidised and crazed.

As the asphalt roof covering cools it gradually hardens to a hard, impermeable, continuous waterproof surface that will have a useful life of 20 years or more. In time the bitumen in the asphalt will become hard and brittle and will no longer be capable of resisting the inevitable movements that occur in any roof covering. It is practice to renew an asphalt roof covering about every 20 years if a watertight covering is to be maintained.

An asphalt roof covering is usually laid to a fall of at least 1 in 40 so that, allowing for deflection under load and constructional errors, there will be a minimum fall of at least 1 in 80 at any point on the roof for rainwater run-off.

There will be considerable fluctuations in the temperature of an asphalt roof surface that is dark coloured and absorbs radiant heat from the sun during the day and cools at night. To reduce temperature change it is practice to dress asphalt with light coloured stone chippings that reflect radiant heat, reduce heat loss at night and ageing fatigue. Stone chippings are bonded to asphalt with a cold applied bitumen solution.

Where an asphalt covered roof is used as balcony, terrace, promenade deck or roof garden, it is practice to add 5% to 10% of grit to the top layer of asphalt to provide a more resistant surface and to use paving grade asphalt to a thickness of 25. For roof gardens three-coat asphalt to a thickness of 30 is used where the surface is not accessible for inspection and repair.

Asphalt is most used as a roof covering on solid decks of concrete and wood wool slabs and less on lightweight decks because of the weight advantage of built-up bitumen and single ply coverings.

At upstands, parapets and curbs to rooflights an upstand skirting at least 150 high is formed in two coats of asphalt 13 thick, with a reinforcing angle fillet of asphalt as illustrated in Fig. 84. The top of the asphalt skirting is turned into a groove cut in a brick joint which is finished with mortar pointing. As an additional weathering, a non-ferrous metal flashing of lead can be dressed down over the skirting from the level of the d.p.c., built into the brick wall. Upstand skirtings to concrete are turned into a groove formed or cut in the concrete and the joint finished with a mortar pointing.

At verges of timber or concrete roofs the asphalt covering is reinforced with expanded metal lathing and run over the edge as a drip with 20 thick asphalt as illustrated in Fig. 84. Curbs to rooflights are either formed in concrete up to which an asphalt skirting is run, or a curb is formed with a pressed metal upstand with timber facing to which an asphalt skirting is formed and reinforced with expanded metal.

Rainwater outlets are usually formed over points of

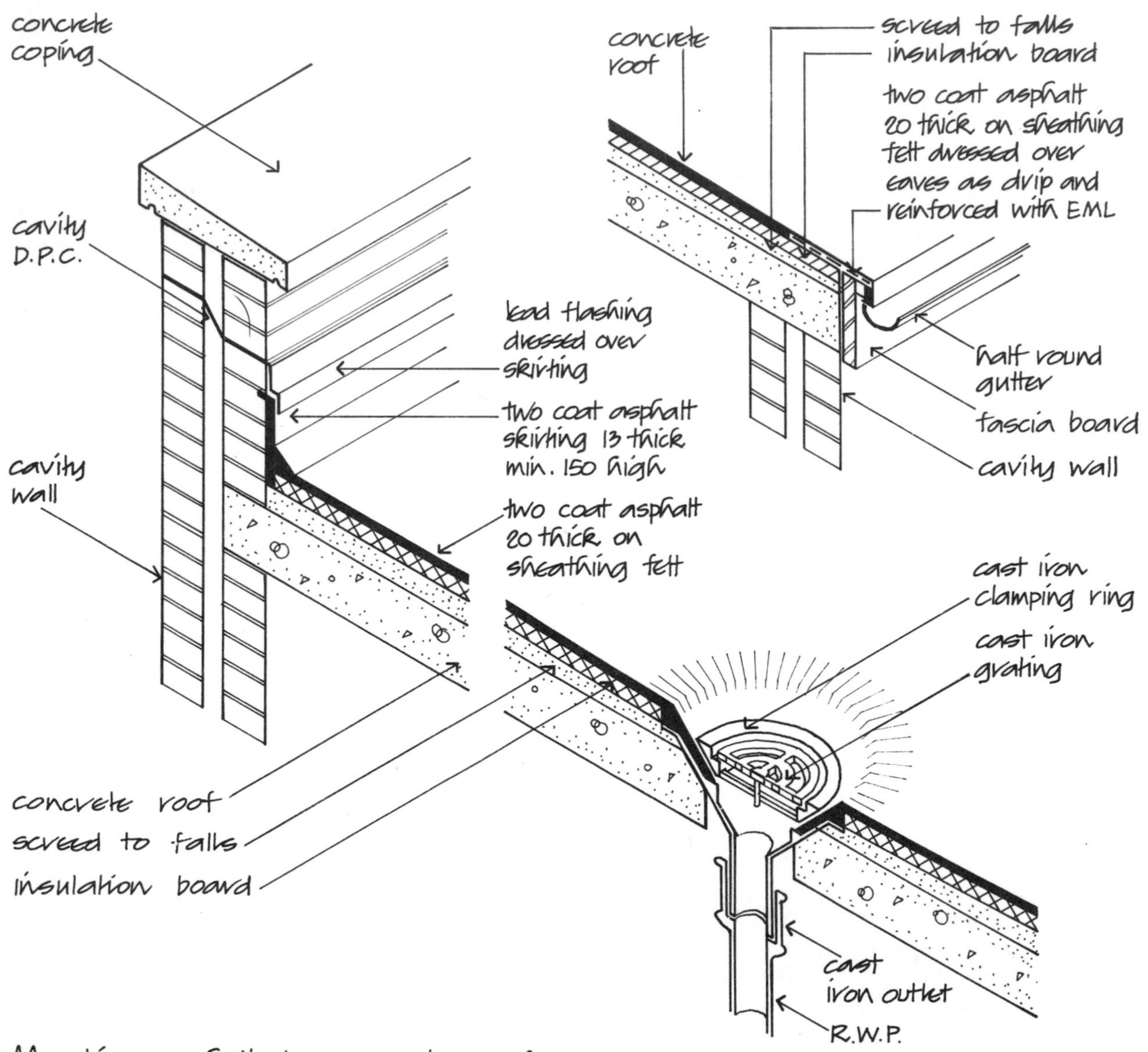

Fig. 84

support of the roof structure so that internal down pipes can be run either against or inside internal columns. The outlets may be constructed in level valleys formed by the fall of the roof. The asphalt is dressed into the rainwater outlet as illustrated in Fig. 84.

Sheet metal coverings The traditional materials that have been used as a weather covering to flat roofs are sheets of lead, copper, aluminium or zinc, laid over flat roof decks. These comparatively small sheets of thin, non-ferrous metal are lapped over and fixed to timber battens fixed to the deck down the slope or fall as illustrated in Volume 1.

The advantage of these materials, particularly lead and copper, is their durability to the extent that they may well have a useful life in excess of that of the majority of buildings. The disadvantage of these materials is that the very considerable labour required in fixing and jointing makes them a comparatively expensive roof covering.

Built-up roof coverings Built-up roof coverings of bitumen impregnated felt have been used for many years as an economic form of flat roof covering. The materials used for built-up roofing are:

Organic fibre-based felts The original felt roofing was made of sheets with a base of felted animal or vegetable fibres satured with bitumen, laid overlapping in two or three layers bonded to the roof deck and together with hot bitumen. The disadvantages of this short life roof covering are that the material is subject to rot due to the absorption of moisture, it may rupture after a few years due to the continuous strain of extension and contraction with temperature changes that are normal to exposed roofs and it has high surface spread of flame characteristics. The use of fibre-based bitumen felt as a roof covering is now confined to use as a shortlife covering for sheds and other temporary buildings.

As a substitute for organic fibre-based felt, asbestos fibre-based felt was used for its better resistance to loss of strength by water absorption and its improved resistance to damage by fire. Because of the hazards to health in the use of asbestos fibre and because of the greater tensile strength of glass fibre-based felts, asbestos fibre felt is no longer used.

Glass fibre-based felt This roof sheeting was first introduced in the 1950s. The glass tissue of this felt is composed of insoluble glass fibres held together with an adhesive. Glass fibre-based felt built-up roofing is the cheapest of the felt roofing materials used. The glass fibre-based felts, which have poor tensile strength and low fatigue endurance in resisting the strain of extension and contraction due to temperature changes, are vulnerable to mechanical and impact damage. This comparatively cheap type of roofing may be used as the first layer in built-up roofing.

High performance roofing

Polyester base roofing Since the 1960s a range of 'high performance polyester roofing' has been produced comprising a polyester fabric with bitumen coating. The high tensile strength and tear resistance of the polyester fabric gives good resistance to the strains of extension and contraction, rupturing and damage.

Two types of bitumen coating are used, oxidised bitumen (SBS) and modified bitumen (APP). The latter coating provides improved fatigue resistance and flexibility. High performance polyester-based roofing, which is more expensive than glass fibre felt, is much used as the under and top coatings to built-up roofing.

The surface of the bitumen coated sheets is coated with fine granules of mineral to prevent the sheets sticking together in the roll. The top surface of the top layer is supplied with mineral granules as a weathering finish.

Attaching built-up roofing to roof decks

Full bonding, partial bonding The first layer of built-up roof sheeting has to be attached to the roof deck surface to resist wind uplift. The method of attachment chosen depends on the nature of the surface of the roof deck. The conventional method of attaching built-up sheet coverings to a roof is by the pour and roll technique, in which hot bitumen is poured onto the deck in front of the roll of sheeting which is then continuously rolled out onto the bitumen. The two methods of bonding that are used are full bonding, where the whole of the bottom layer of sheeting is attached to the deck, and partial bonding.

Partial bonding is used on roof deck surfaces with high laminar strength, e.g. wood wool, polyurethane and polyisocyanuate which will provide adequate attachment by partial adhesion and allow some movement of the sheeting independent of the deck. Full bonding is used on surfaces of low laminar strength such as mineral fibre and expanded polystyrene.

For partial bonding a first layer of perforated glass fibre underlay is laid on the deck. This layer is fully bonded to the deck around the perimeter of the roof and around openings for a width of 500 to prevent wind uplift of edges. The next layer of polyester base roofing is then fully bonded to the first layer by the pour and roll method and the third layer of polyester base sheeting is fully bonded to the second layer.

Where the fully bonded method of attachment is followed, two layers of polyester base sheeting are generally used.

Joints The joints between the sheets of roofing used in built-up roofing are overlapped 50 along the long edges of the sheets, down the slope or fall of the roof and 75 at end laps. The joints of sheets in successive layers are staggered so that there is not too great a build-up of thickness of material on the roof. The overlap joints are fully bonded in hot bitumen.

Skirtings and upstands At upstands to parapets, abutments and curbs to rooflights an upstand skirting at least 150 high is formed. To take out the sharp right-angled junction of the roof and upstand, a timber, wood fibre board or polyurethane angle fillet, 50 × 50, is fixed so that the covering is not damaged by being turned up at a right angle. The first layer of three-ply roofing is finished at the angle fillet, the next two layers in three-ply work and both layers in two-ply work are dressed up as a skirting as illustrated in Fig. 85. The top edge of the skirting is covered with a lead flashing, the top edge of which is wedged into a groove in a brick joint.

Eaves and verges At eaves and verges a welted drip is formed with a strip of mineral faced roofing. The strip of roofing is nailed to a batten or fascia board, bent to form a 75 deep drip edge and then turned onto the roof and bonded between the two top layers of sheeting as illustrated in Fig. 86. At verges an edge upstand is

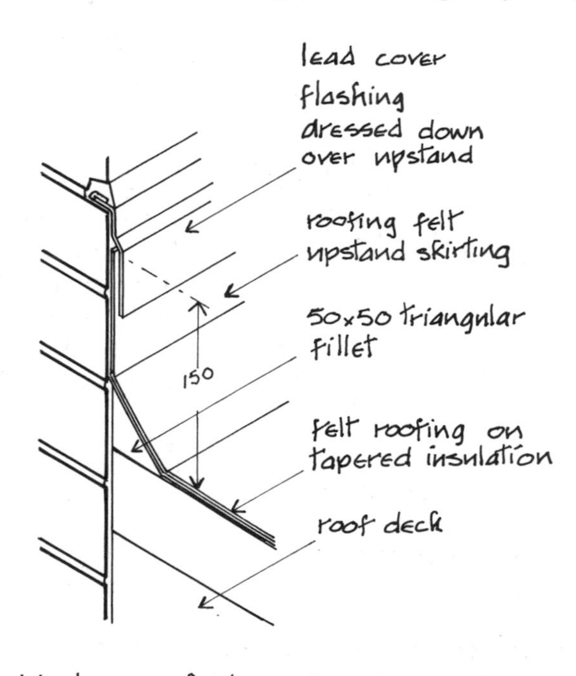

Fig. 85

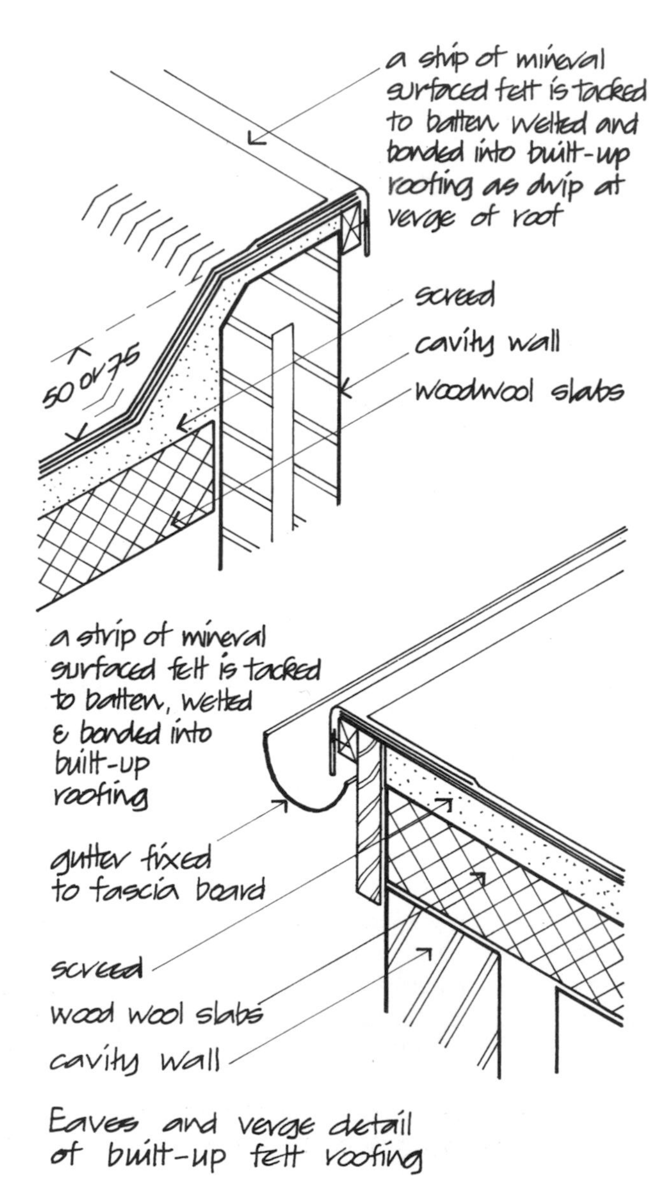

Fig. 86

formed with timber or in the roof surface to prevent water running off the roof. A strip of mineral surfaced roofing is nailed to a batten or fascia board, turned up to form a welt and bonded into the two top layers of roofing as illustrated in Fig. 86.

Expansion joint Where there is an expansion joint formed in the structure to accommodate anticipated thermal, structural or moisture movements, it is necessary to make some form of upstand in the roof each side of the joint. The roofing is dressed up each side of the joint as a skirting to the upstands. A plastic coated metal capping secured with secret fixings is used as weather capping to the joint as illustrated in Fig. 87.

Torching On roofs where access is difficult for equipment for heating bitumen and also for surfacing existing coverings, it is convenient to use a felt coated with additional bitumen on one side which can be heated by a torch for adhesion to the roof surface onto which the felt is rolled either as built-up roofing on new roofs or as one layer on old roof coverings. Polyester

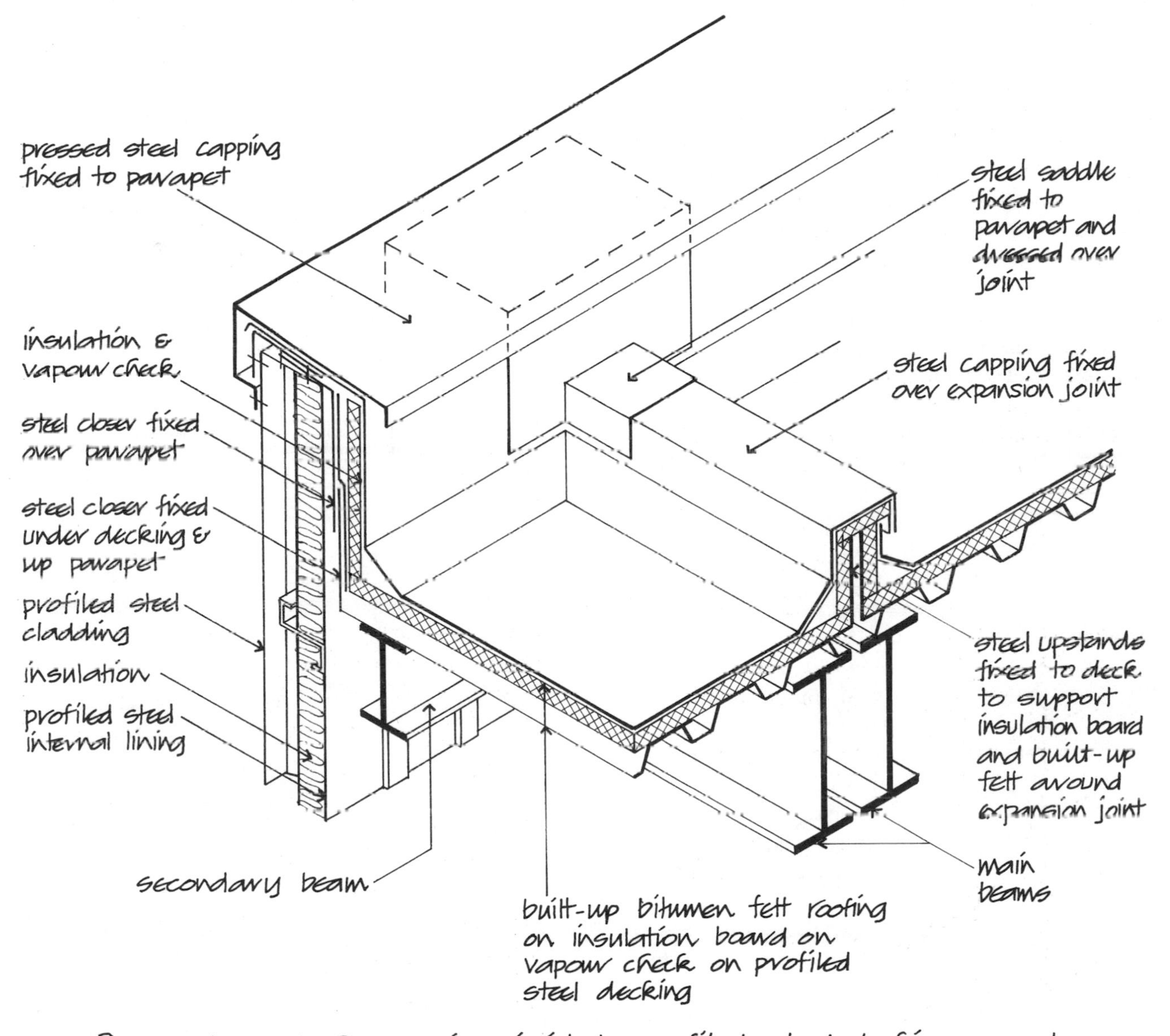

Fig. 87

base sheeting is used for torched attached sheeting. The face of the sheeting is heated by the flame of a gas torch causing the bitumen to soften and bond to the roof as the sheet is progressively unrolled onto the roof.

Nailing The first layer of roofing of polyester or glass polyester-based sheeting is nailed to timber boarded decks with galvanised nails at 200 centres. The second and third layers are then attached by full bonding by the pour and roll method with 50 overlaps at long edges and 75 at end laps.

Single ply roofing

Since the mid 1960s a range of polymeric single ply roofing materials has been produced. Single ply roofing materials are very extensively used for flat roofing in the US and to a considerable extent in Northern Europe. These materials provide a tough, flexible, durable lightweight weathering membrane which is able to accommodate thermal movements without fatigue. To take the maximum advantage of the flexibility and elasticity of these membranes the material should be loose laid over roofs so that it is free to expand and contract independently of the roof deck. To resist wind uplift the membrane is held down either by loose ballast, a system of mechanical fasteners or adhesives.

The materials used in the manufacture of single ply membranes may be grouped as thermoplastic, plastic elastic and elastomeric.

Thermoplastic These materials include:

PVC — Polyvinylchloride is a tough material with good flexibility, resistance to fire and oil damage that can be solvent or heat welded.

CPE — Chlorinated polyethylene is a tough material with good flexibility and resistance to fire and oil that can be solvent or heat welded.

CSM — Chlorosulphonated polyethylene is similar to CPE except that the material on exposure to solar radiation stiffens to produce greater toughness and elasticity. It can be solvent or heat welded.

VET — Vinyl ethylene terpolymer is a mixture of ethyl vinyl acetate, terpolymer and PVC which has long term flexibility, good flame and spread resistance and can be solvent or heat welded.

Plastic elastic These materials include:

PIB — Polyisobutylene has good resistance to chemicals and oxidation, flexibility and can be solvent or heat welded.

IIR — Butyl rubber is highly elastic, does not heat soften but is susceptible to exposure to ozone. It cannot be heat welded and is jointed with adhesive.

Elastomeric EPDM – Ethylene propylene diene monomer – materials are flexible and elastic with good resistance to oxidation, ozone and ultraviolet degredation. Such materials are jointed with adhesives.

These single-ply materials are impermeable to water, moderately permeable to moisture vapour, flexible and maintain their useful characteristics over a wider range of temperatures than the materials used for built-up roofing. To enhance tear resistance and strength, these materials may be reinforced with polyester or glass fibre fabric.

The cheapest and most used material, PVC, is made in sheet thicknesses of 1.2 mm to 1.5 mm.

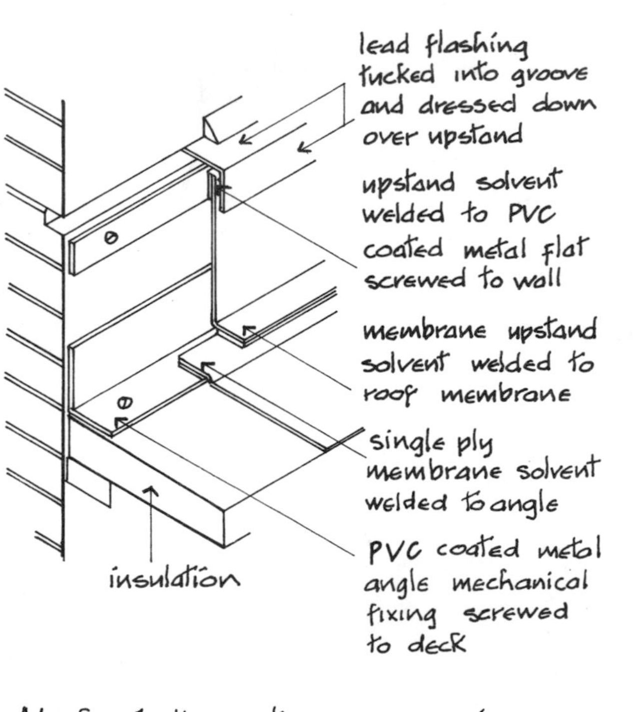

Fig. 88

Jointing single-ply membranes The joints between the sheets of the membrane are sealed either by solvent or hot air welding the 50 minimum overlap between thermoplastic sheets or by adhesive for elastomeric sheets. Solvent welding is the method most used for thermoplastic sheets. As the solvent is spread between the sheets, pressure is applied to the joint by dragging a sand-filled bag over the top of the joint. As the solvent evaporates to air, the edges of the two sheets at the overlap become welded together to make a watertight joint.

Hot air welding is mainly used to make joints at corners and upstands where a solvent welded joint might be difficult to make. The heated nozzle of the welding machine is inserted into and run along the joint. As the material softens, pressure is applied over the seam with a silicon rubber roller to make firm contact of the two sheet edges to form a watertight joint. The 50 overlap joint between elastomeric sheet membrane is made watertight by the application of adhesive and tapes that bond and seal the joints.

Laying single-ply membranes The three systems of laying single-ply membranes are:

(1) loose laid membrane
(2) loose laid ballasted
(3) adhered.

Loose laid membranes, which allow maximum

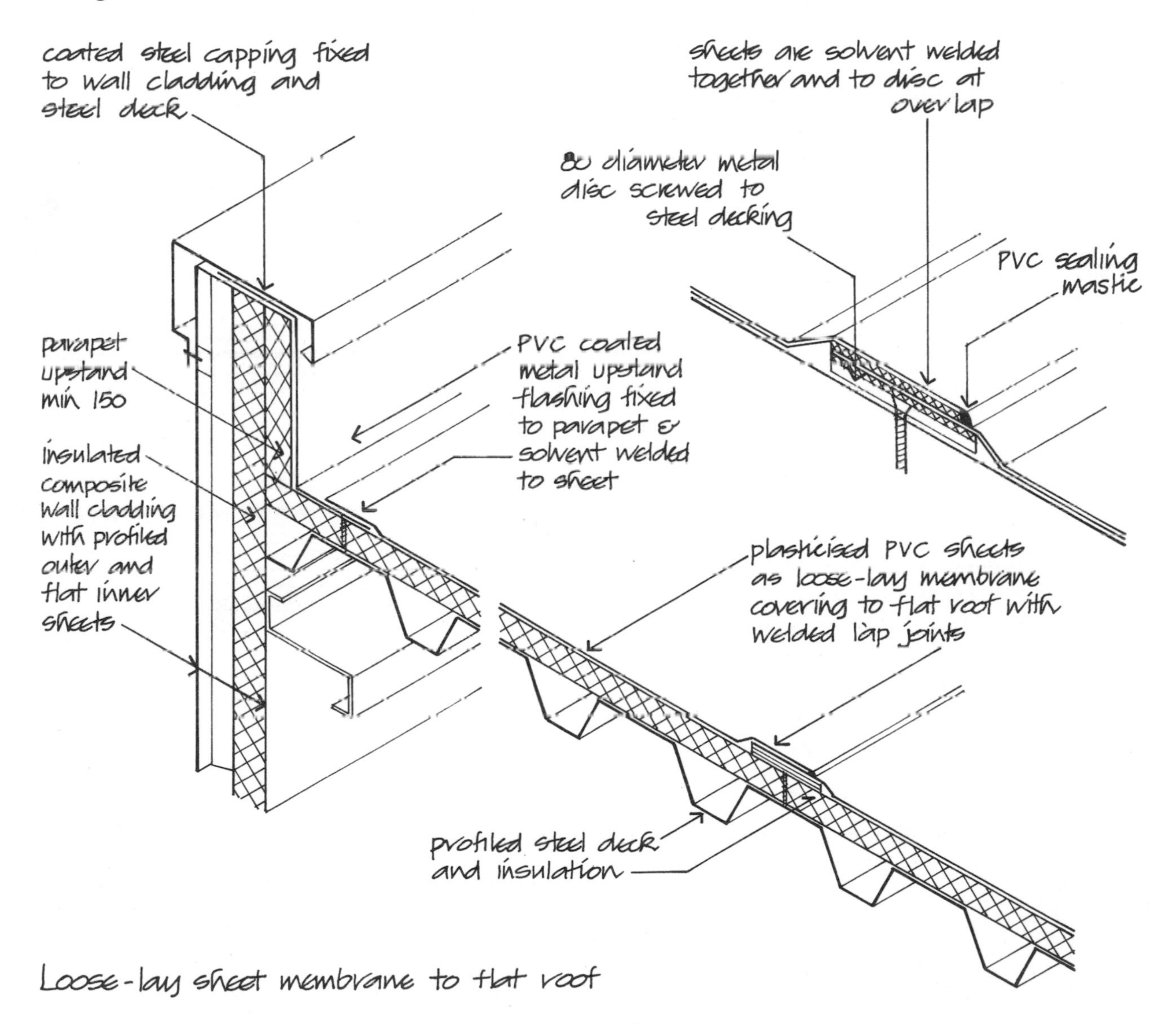

Fig. 89

freedom of movement of the membrane independent of the roof deck, have to be mechanically fastened, both with edge fasteners and fasteners at joints between sheets to provide restraint against wind uplift of this comparatively lightweight material.

Loose laid ballasted membranes employ ballast to weight down the membrane against wind uplift.

Membranes are adhered to the deck or roof against wind uplift where a system of fasteners or ballasting cannot be used.

Mechanical fasteners Both loose laid and loose laid ballasted systems of laying single-ply membranes employ mechanical fasteners to the periphery of the roof and around penetrations of roof covering against wind suction uplift. These fasteners are formed from PVC coated metal angles and flats that are screwed to the deck and upstands. The single-ply membrane is solvent-welded to the fasteners to mechanically secure the membrane against wind uplift as illustrated in Fig. 88. Where single-ply membranes are not ballasted against wind uplift as, for example, on lightweight metal decking that is not designed to carry the load of ballast, it is generally necessary to secure the membrane to the deck by a system of mechanical fasteners. These fasteners take the form of PVC coated metal strips screwed to the deck below the joints between the sheets or a series of flat or round disc washers. One edge of the membrane is fixed under and the other over the flats or washers and the joint between the sheets is solvent-welded or one sheet is solvent-welded to the washers and the other solvent-welded over it as illustrated in Fig. 89.

Loose laid ballasted system The loose laid membrane, which is mechanically secured at the perimeter and penetrations, is weighted with a topping of clean, round, screen size 16 to 32 ballast spread over the membrane to a thickness of at least 50, on a layer of protective fleece that is laid over the membrane. Where pedestrian traffic to the roof is allowed for, paving slabs are laid on a protective fleece over the membrane with a margin of ballast around the roof to allow for movement. With inverted roof systems the ballast or paviors are laid on a filter layer on the insulation that is laid over the single play membrane.

Figure 90 is an illustration of a ballasted membrane to an inverted roof.

Adhered system On sloping and curved roof structures where ballasted or mechanically fastened membranes may not be either practical or effective, it is necessary to secure the single-ply membrane with partial or full bonding to the roof surface depending on the anticipated extent of wind uplift. The membrane is secured with adhesive to the roof surface and the joints are made with either solvent or heat welding.

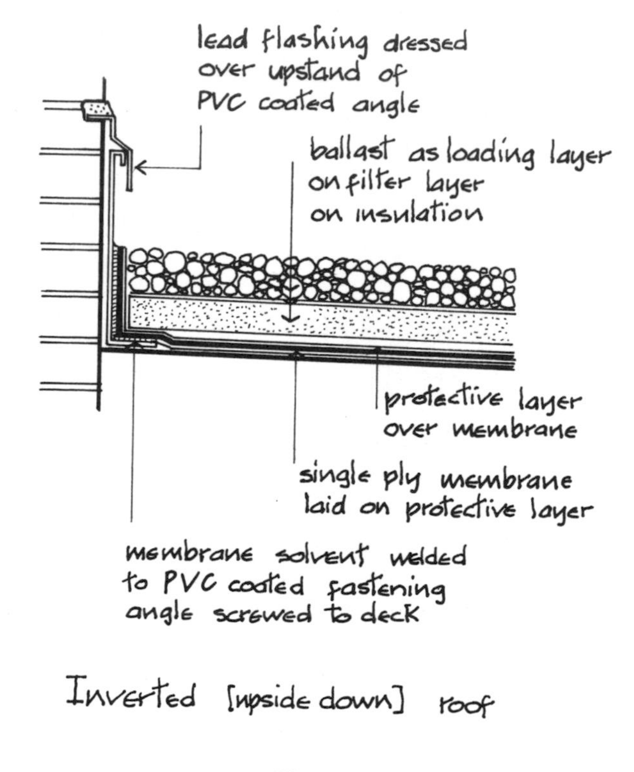

Fig. 90

CHAPTER THREE

ROOFLIGHTS

The useful penetration of daylight through windows in the walls of buildings is some 6.0 to 9.0 inside the building, depending on the level of the head of windows above floor level. It is plain, therefore, that inside a building in excess of 12.0 to 18.0 in width or length there will be areas to which a useful degree of daylight will not penetrate through windows in walls. The traditional means of providing daylight penetration to the working surfaces of large single-storey buildings is through rooflights either fixed in the slope of roofs or as upstand lights in flat roofs. Usual practice was to cover the middle third of each slope of symmetrical pitch roofs and the whole of north facing slopes of north light roofs with rooflights and about a third of flat roofs with upstand rooflights. The penetration of daylight through these rooflights provided adequate natural light on working surfaces in factories and workshops for bench level, hand-operated activities.

With the very considerable increase in automated processes of manufacturing not dependent on bench level hand-controlled operations, the provision of daylight through roofs is often a disadvantage due to the variability of natural daylight and shadows cast by heavy overhead moving machinery, to the extent that artificial illumination by itself has become common in many modern factories and workshops.

The advantage of economy and convenience in the use of natural lighting from rooflights has to be balanced against the disadvantages of poor thermal and sound insulation, discomfort from glare, solar heat gain and the hazards of fire from the materials used for rooflights.

FUNCTIONAL REQUIREMENTS

The primary function of rooflights is:

Admission of daylight.

The functional requirements of rooflights as a component part of roofs are:

Strength and stability
Resistance to weather
Durability and freedom from maintenance
Fire safety
Resistance to the passage of heat
Resistance to the passage of sound
Security.

Daylight

The prime function of rooflights is to admit an adequate quantity of daylight with minimum diversity and without excessive direct view of the sky or penetration of direct sunlight. The area of rooflights chosen for single-storey buildings is a compromise between the provision of adequate daylight and the need to limit loss of heat through the lights. The ratio of the area of rooflights to floor area is up to 1 to 6 for most factory buildings with pitched roofs and up to 1 to 3 for roofs with vertical monitor lights.

The quantity of daylight that is admitted through a window or rooflight is expressed as a daylight factor. This is defined as the ratio of the daylight illumination at a point on a plane due to the light received directly or indirectly from a sky of assumed or known luminance distribution, to the illuminance on a horizontal plane due to an unobstructed hemisphere of this sky.

It has been practice for some time to determine the required area of window or rooflight from the minimum daylight factor, which is the daylight at the worst lit point on the working plane assuming that if there is adequate daylight at the worst lit point there will be adequate light at all other points. This assumption is reasonable where specific critical judgements in daylight alone, which depend on colour, contrast and detail, such as clinical judgements in a hospital, have to be made. Even though daylight at the worst lit point may be satisfactory for a specific task, the daylight in the room or area as a whole may not be entirely satisfactory. The minimum daylight factor that has been in use, which gives an indication of daylight requirements at a given point for the performance of critical tasks, does not provide a wholly satisfactory indication of the distribution of daylight in a particular room or area for a variety of activities.

In DD73.1982, which is a draft for development as a preliminary to the publication of a new Code for Daylighting, it is proposed that an average daylight factor and the uniformity ratio be adopted as a better indication of the daylight requirements more closely matching the specific requirements of distribution of illuminance and luminance in a specific room than that given by a minimum daylight factor alone. It is likely that this proposal will be accepted for inclusion in a new Code for Daylighting.

The average daylight factor may be calculated from the formula:

$$\text{average daylight factor} = \frac{\text{total incident light flux on working plane}}{\text{outdoor illuminance} \times \text{area of working plane}}$$

The uniformity ratio, which is an indication of the diversity of daylight, is the ratio of

$$\frac{\text{minimum daylight factor}}{\text{average daylight factor}}$$

The lower the uniformity ratio the greater the diversity of illuminance, and the higher the uniformity ratio, the better the lighting.

The recommendations for average daylight factors and minimum uniformity ratios of daylight for working places given in DD73.1982 are:

- Average daylight factors on reference plane, 5 for full daylight and 2 for supplemented daylight
- Uniformity ratio, 0.7 for top lit, 0.3 for side lit with full daylight and supplemented daylight.

Glare, discomfort and disability Glare is the word used to describe the effect of excessive contrast, in a fixed direction of view, between a very bright light source and a relatively dark background light, such as the contrast of a window at the end of a dark corridor which may cause discomfort without reducing vision, or the contrast of a very bright skylight at the top of a stair which can cause disability to vision.

Both discomfort and disability glare are caused by excessive contrast and unfavourable distribution of luminance in the visual field. Discomfort glare, which is directly related to the absolute luminance irrespective of whether there is unfavourable contrast present, may occur on a bright day in a room that is comfortable on a dull day.

The discomfort glare index is an expression of subjective discomfort in relation to the luminance of windows, average luminance of interiors and the direction of the source of glare related to the normal direction of viewing. A table of glare constants is set out in DD73.1982 to limit the degree of discomfort due to glare which is generally acceptable to occupants of rooms. Rooflights should be of sufficient area to provide satisfactory daylight and be spaced to give reasonable uniformity of lighting on the working surface without excessive direct view of the sky, to minimise glare or penetration of direct sunlight and to avoid excessive solar heat gain. In pitched roofs, rooflights are usually formed in the slopes of the roof to give an area of up to one sixth of the floor area and spaced as indicated in Fig. 91 to give good uniformity and distribution of light. Rooflights in flat roofs are constructed with upstand curbs to provide an upstand to which the roof covering can be finished. The area of the lights should be up to one sixth of the floor area and spaced similar to lights in pitched roofs. Monitor rooflights with either vertical or sloping sides give reasonable uniformity and distribution of light with the spacing shown in Fig. 91. Vertical monitor lights should have an area of up to one third of the floor area to provide adequate daylight to the working plane. North light roofs, with the whole or a large part of the north facing slopes glazed, are adopted to avoid or minimise sun glare and solar heat gain. In consequence the daylight is from one direction only with less uniformity of distribution and stronger modelling than with lights in symmetrical pitch or flat roofs. The area of glazing of north lights should be up to one third of the floor area and the spacing of lights is shown in Fig. 91.

Strength and stability

The materials used for rooflights, glass and flat or profiled, transparent or translucent sheets, are used in the form of thin sheets to obtain the maximum transmission of light and for economy. Glass which has poor tensile strength requires support at the comparatively close centres of about 600 to provide adequate strength and stiffness as part of the roof covering. Plastic sheets that are profiled to match sheet metal and fibre cement roof coverings, for the sake of weathering at end and side laps, have less strength in the material of the sheets and stiffness in the depth of the profiles than steel sheets and generally require support at closer

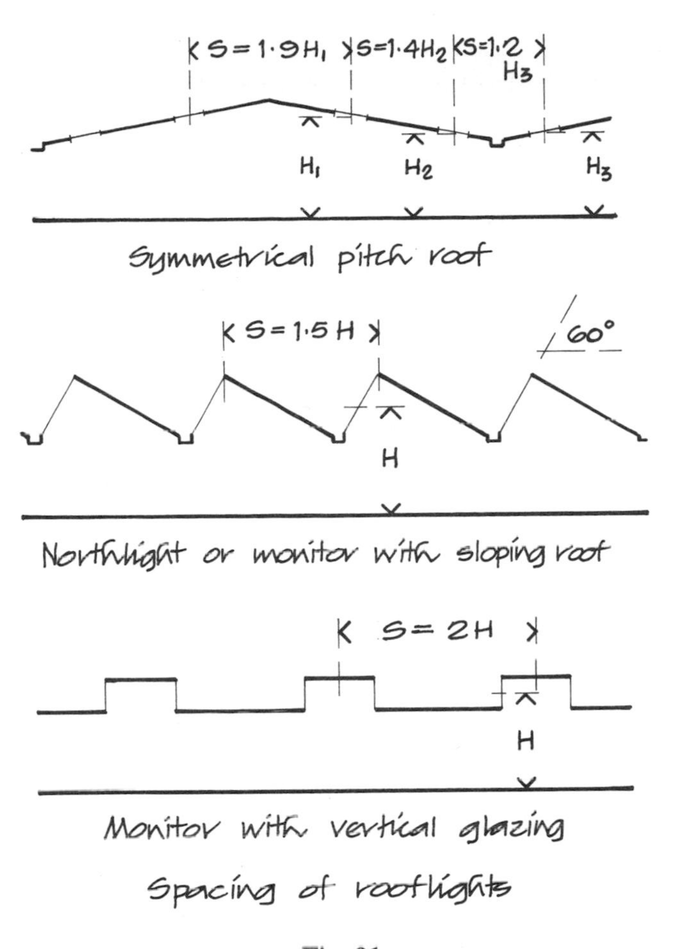

Fig. 91

centres. Plastic sheets that are extruded in the form of double and treble skin cellular flat sheets have good strength and stiffness in the nature of the material and the cellular form in which they are used.

Resistance to weather

The metal glazing bars used to provide support for glass are made with either non-ferrous flashings or plastic cappings and gaskets that fit over the glass to exclude wind and rain, together with top and bottom non-ferrous flashings to overlap and underlap profiled sheet coverings pitched at least 15° to the horizontal.

Profiled plastic sheets fixed in the slopes of pitched roofs are designed to provide an adequate side lap and sufficient end lap to give the same resistance to the penetration of wind and rain as the profiled metal cladding in which they are fixed at a pitch or slope of not less than 10° to the horizontal.

Where profiled plastic sheets are fixed as rooflights in roof slopes of less than 10° to the horizontal it is necessary to seal both side and end laps to profiled metal sheeting with either double-sided tape or a silicone sealant to exclude wind and rain.

Profiled plastic sheets fixed as rooflights in roof slopes covered with fibre cement sheets do not make a close fit to either the side or end laps to the fibre cement sheets. Because of the considerable gaps between the sheets it is difficult to form a seal to exclude wind and rain.

Cellular flat plastic sheets are fitted with metal or plastic gaskets to weather the joints between the sheets fixed down the slope of roofs and non-ferrous metal flashings at overlaps at the top and bottom of sheets.

Rooflights in flat roofs and low pitch roofs are fixed on a curb to which an upstand skirting of the roof covering is dressed to exclude wind and rain with the rooflights overlapping the curb to weather rain, with wind stops as necessary between the light and the curb.

Durability and freedom from maintenance

Unlike the plastic materials that are used in rooflights, glass does not suffer discoloration and yellowing with age and maintains its bright, lustrous, fire-glazed finish for the useful life of buildings. Because of its smooth, hard finish glass is easily cleaned by washing to maintain its bright lustrous finish. Plastic materials more readily dirt stain than glass due to the surface of these materials and cannot effectively be cleaned to return the material to its initial clean finish.

Fire safety

The requirements from Part B of Schedule 1 to the Building Regulations 1991, Fire Safety, that are relevant to rooflights are those that limit internal fire spread (linings) and those that limit external fire spread.

To limit the spread of fire over the surface of materials, the guidance to Approved Document B limits the use of materials that encourage the spread of flames across their surfaces when subject to intense radiant heat and those that give off appreciable heat when burning. The use of thermoplastic materials in rooflights and lighting diffusers is limited by reference to the classification of the lower surface of the material related to its spread of flame characteristics and by limitations to the size and disposition of rooflights and lighting diffusers related to roof or ceiling areas.

To limit the spread of fire between buildings the guidance in Approved Document B sets limits on the use of materials used in roof coverings and rooflights that may encourage the spread of fire. A minimum distance from boundaries is related to the spread of flame characteristics of materials used in roof coverings which limits the use of thermoplastic materials in rooflights. The separate limitations in the use of plastic in rooflights relates to the classification of surface spread of flame of the material to distances from boundaries and sets limits on the spacing and size of plastic rooflights.

Resistance to the passage of heat

Because the thin sheets of glass or plastic used on rooflights offer little resistance to the transfer of heat, the area of rooflights is limited. Some reduction in the transfer of heat is effected by the use of double-glazing in the form of two skins of glass or plastic sheeting and the double or triple skin, cellular flat sheets of plastic. It is plainly impractical to clean the surfaces of glass or plastic sheets inside the air space of double-glazing which may in time suffer a reduction in light transmission due to the accumulations of dust and condensation on surfaces inside the air space. Sealed double-glazing units of glass or plastic sheets will prevent this loss of light transmission.

The practical guidance given in Approved Document L to meeting the requirements of Part L of Schedule 1 to the Buildings Regulations 1991, for the conservation of fuel and power, sets limits to the maximum U value of roofs at 0.45 W/m^{2}K, and maximum areas of single-glazed areas of rooflights for all buildings other than dwellings, at 20% of roof area. In the notes of guidance to the maximum area of rooflights is a statement that where double-glazing is used, the maximum glazed area may be doubled and where double glazing is coated with low emissivity coating, and where triple-glazing is used, the maximum glazed area may be trebled.

Resistance to the passage of sound

The thin sheets of glass and plastic used in rooflights offer little resistance to the transfer of sound. Double skin glass and plastic sheet will result in minimal reduction in sound transfer. Where sound reduction is a critical requirement of a roof it is necessary to use the mass of a material such as concrete for the roof without any rooflights or with concrete lens lights.

Security

Single-storey buildings clad with lightweight metal cladding to roofs and walls are as vulnerable to forced entry through windows, doors, thin wall and roof cladding and both glass and plastic rooflights. There is little point, therefore, in seeking to secure windows and doors when the surrounding wall can as easily be broken through. Unattended buildings in isolated situations are just as likely to suffer damage by vandalism to any one of the flimsy or brittle materials of the fabric. Security against forced entry and vandalism is best achieved by reasonably secured perimeter fencing and effective day and night surveillance.

MATERIALS USED FOR ROOFLIGHTS

Before fibre cement and metal sheeting came into use the traditional material for rooflights was glass laid in continuous bays across the slopes of roofs and lapped under and over the traditional roofing materials, slate and tile, down the slope. The majority of rooflights today are of translucent profiled sheets of plastic formed to the same profile as the roof sheeting. For economy in material and simplicity of fixing, profiled translucent sheeting is extensively used instead of glass.

Glass

Glass is used in the form of flat sheets supported by metal glazing bars fixed in the slope of roofs of buildings, used as flat sheets fixed in metal frames for deck and lantern lights and shaped for use as domelights.

The types of glass used for rooflights today are float glass (see Volume 2) that is transparent and has flat, parallel, bright, fire-polished surfaces with little distortion, solar control glass to limit the admission of solar radiation, patterned glass which is textured or patterned and is translucent, and wired glass which is used to minimise the danger from broken glass during fires.

Glass has poor mechanical strength and requires the support of metal glazing bars at comparatively close centres of about 600 for use as patent glazing in the slope of roofs and as side wall glazing. The need for glazing bars to provide support for glass and the necessary flashings of caps and gaskets for weathering adds considerably to the cost of glass for rooflights. Glass affords little resistance to the transfer of heat, the U value of single-glazing being 5.7 W/m^{2}K for 6 glass. A comparatively small increase in the resistance to the transfer of heat is effected by the use of double-glazing, the U value of double-glazing being 2.8 W/m^{2}K. The thin solid material of glass offers poor resistance to the transfer of sound.

When subjected to the heat generated by fires in buildings ordinary glass quickly cracks, falls away from its support and presents a hazard to those below. It is also a poor barrier to the spread of fire within and between buildings. The wire mesh that is embedded in wired glass will hold together glass that cracks when subjected to the heat of fires and so minimise the danger from broken glass. It will also maintain the glass as a barrier to the spread of fire for a period of from a half to one hour. For these reasons wired glass is often used in rooflights. The principal advantage of glass is that as it is transparent it provides a clear undistorted view and also because it has a bright, fire-glazed finish it is easily cleaned by washing and maintains its clear lustrous finish without discolouring and yellowing with age. Glass is still used for roof and wall lights in the form of patent glazing for the sake of a clear view and its lustrous finish, compared to the translucent, dull opaque finish of the cheaper material, plastic.

Profiled, cellular and flat plastic sheets

Plastic sheet material which is transparent or translucent and can be shaped to match the profiles of metal and fibre cement sheets is extensively used as rooflights in the slopes of roofs and on flat roofs in the form of lay lights and domelights.

The materials most used for profiled sheeting are:

(1) uPVC-polyvinyl chloride – rigid PVC
uPVC is the cheapest of the translucent plastic materials used for rooflights. It has reasonable light transmittance (77%), reasonable impact and scratch resistance, adequate strength for use as a profiled sheet for roofing and good resistance to damage in handling, fixing and use. It is resistant to attack from most chemicals and has a useful life of 20 years or more. On exposure to solar radiation the material discolours to the extent that there is appreciable yellowing and reduction of light transmission after some 10 years. Due to its low softening point the material, when subjected to the temperatures generated by fires, softens but does not readily burn.

(2) GRP – glass reinforced polyester
GRP has very good impact resistance, rigidity, dimensional stability and fairly good scratch resistance. The material is translucent and has moderate to reasonable light transmittance of from 50% to 70%. GRP has very good durability and resistance to damage in handling, fixing and use. When subjected to the temperatures generated by fires GRP is usually inflammable.

The materials most used for flat sheet rooflights, laylights and domelights are:

(1) PC – polycarbonate
This material is used as extruded sheet in single, double and triple skin rooflights both in the flat sheet and the cellular forms. It has good light transmittance (up to 88%), good resistance to weathering, reasonable durability and extremely good impact resistance (for this characteristic it is sometimes referred to as 'shatterproof'). Polycarbonate is the most expensive of the plastic materials used for rooflights and is used principally for its impact resistance in situations where glass and other plastics would be damaged and for the improved resistance to transfer of heat of the double and triple skin forms.

(2) PMMA – polymethyl methacrylate
This plastic is used for shaped rooflights. It has a hard smooth finish that is particularly free from dirt staining, has excellent chemical resistance, good impact resistance and good resistance to ultraviolet radiation. The material softens and burns readily when subject to the heat generated by fires.

ROOFLIGHTS

Rooflights in pitched roofs covered with slate or tile take the form of continuous bays of glass fixed in the slope, and rooflights in pitched roofs covered with profiled sheeting take the form of bays of profiled sheeting to match the profile of the roofing. The whole or major part of north light roofs and the upstands of monitor rooflights are generally finished with glass or flat polycarbonate sheets as illustrated in Fig. 92.

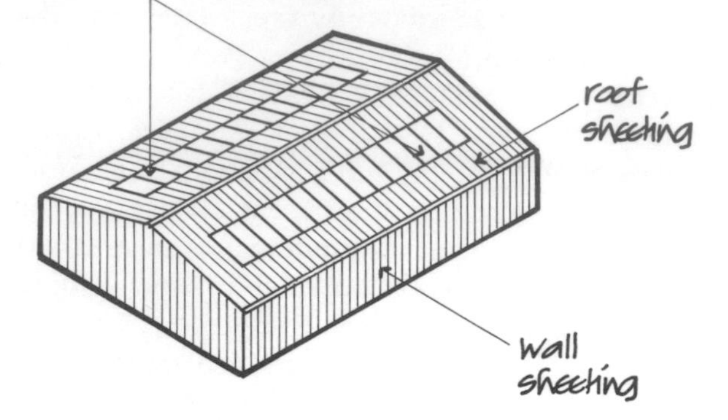

Roof lighting to symmetrical pitch roof

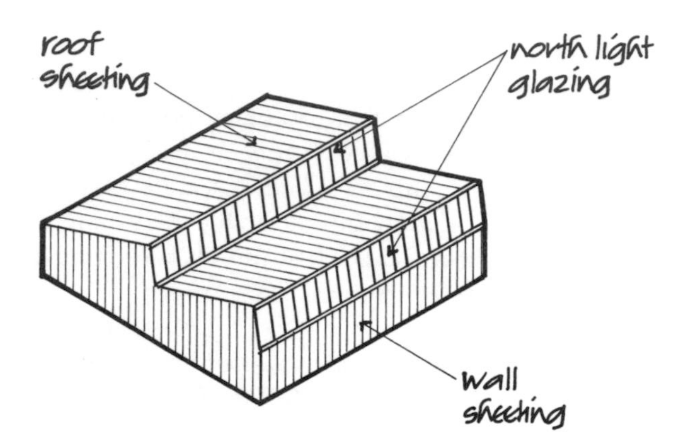

North light roof glazing

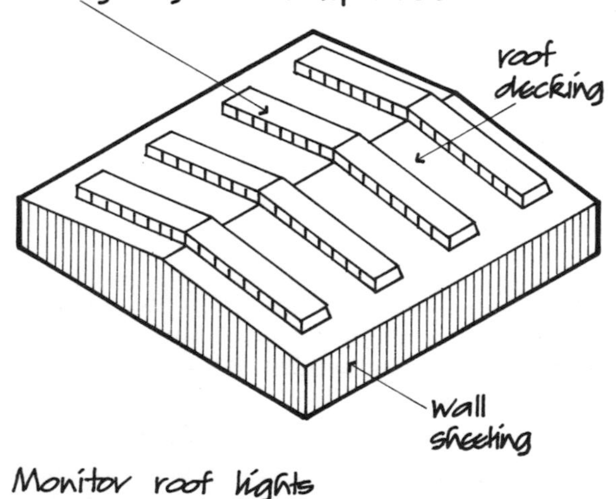

Monitor roof lights

Fig. 92

The most straightforward way of constructing rooflights in pitched roofs covered with profiled sheeting is by the use of sheets of uPVC or GRP formed to match the profiles of the roof sheeting. The profile of translucent sheeting does not generally closely match the profile of metal roof sheeting so that it is impossible to achieve a close fit between the translucent and metal sheeting over wide bays of rooflights. To minimise the mismatch of profiles it is necessary to limit the width of rooflights to comparatively narrow widths, particularly on low pitch roofs where sealed end laps are required to exclude wind and rain as illustrated in Fig. 93.

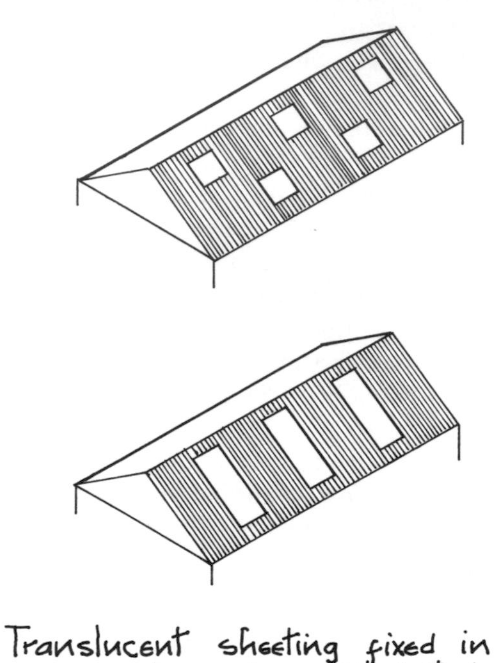

Translucent sheeting fixed in narrow bays to profiled steel sheets

Fig. 93

Translucent glass-reinforced polyester sheets composed of thermosetting polyester resins, curing agents, light stabilisers, flame retardants and reinforcing glass fibres are roll formed in a range of profiles to match most profiled metal and fibre cement sheets. The light transmission of the clear sheets is 70%. Three grades of GRP sheet are produced to satisfy the conditions for external fire exposure and the surface spread of fires set out in the Building Regulations. The material has good strength and stability, is lightweight and shatterproof and has a life expectancy of 20 to 30 years. The sheets offer poor resistance to the transfer of heat, single skin sheets having a U value of 5.7 W/m^{2}K and double skin sheets 2.8 W/m^{2}K.

The sheets are laid to match the abutting metal or fibre cement cladding sheets with side overlaps of profiles to adjacent sheets and under and over end laps to match those of the surrounding sheeting. Sheets with a profile that is less than 35 deep should not be used on a roof pitched at less than 10° and those with profiles 35 or more deep can be laid on roofs pitched as low as 6° to the horizontal provided all laps are sealed.

All side laps between translucent sheets and between translucent sheets and metal and fibre cement sheets should be sealed with self-adhesive closed cell PVC sealing tape to make a weathertight joint. End laps between translucent sheets and between translucent sheets and metal and fibre cement sheets to roofs pitched below 20° should be sealed with beads of silicone sealant. In common with other lightweight sheeting material used for roofing, the fixing of these sheets is critical to resist uplift from wind suction which dictates the necessary centres for the fixing of fasteners. The sheets are fixed with the same type of fasteners that are used for metal or fibre cement sheeting, self-drilling and self-tapping fasteners to metal purlins or spacers and hook bolts to steel purlins with PVC washers and neoprene gaskets to make a weathertight seal. Usual practice is to secure the sheets by fasteners driven through the trough of corrugations or profiles into purlins or spacers. Stitching fasteners are driven through the crown of sheets at side and end laps.

Double skin rooflights are constructed with two skins of GRP as illustrated in Fig. 94, with two sheets of the same profile as the metal or fibre cement sheet roof covering or one to match the cladding and the lower to match the profile of lining sheets. Profiled, high density foam spacers, bedded top and bottom in silicone mastic

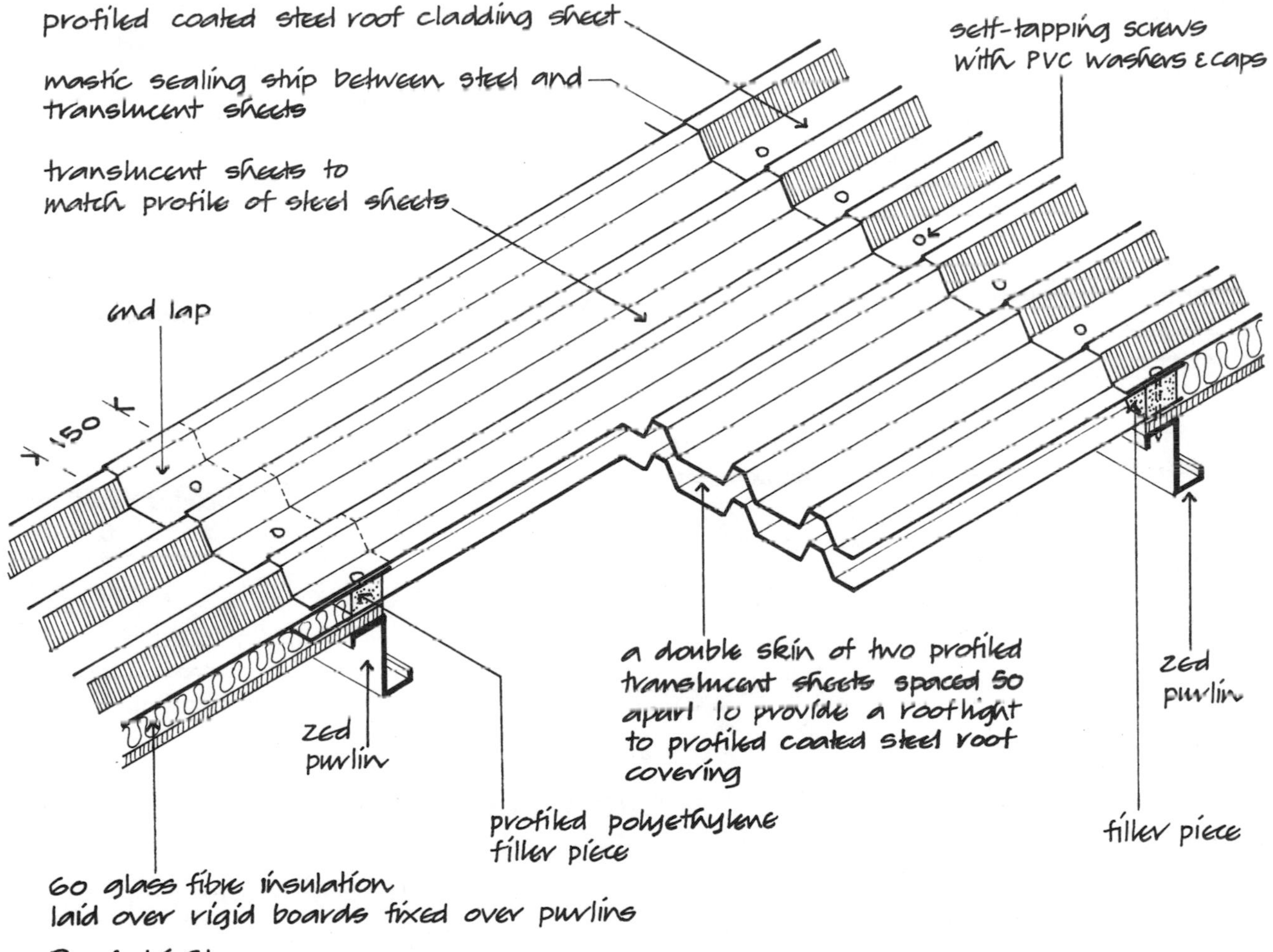

Fig. 94

are fitted between the sheets to maintain the air space and seal the cavity. Double-sided adhesive tape is fixed to all side laps of both top and bottom sheets as a seal. The double skin rooflight is secured with fasteners driven through the sheets and foam spacers to purlins. Stitching screws are driven through the crown of profiles at side and end laps.

Factory-formed sealed double skin GRP rooflight units are made from a profiled top sheet and a flat undersheet with a spacer and sealer.

Translucent polyvinyl chloride (uPVC) sheets are produced in a range of profiles to match most metal and fibre cement sheeting. The material has good impact resistance, reasonable strength and stability and is lightweight and shatterproof. It has a life expectancy of ten years or more because, even though the material is ultraviolet stabilised, it will gradually discolour and lose transparency to an appreciable extent. The sheets provide poor thermal insulation; the *U* value of single skin sheeting is 5.7 W/m^{2}K and double skin 2.7 W/m^{2}K. Because of the low softening point of the material, uPVC sheets soften but do not readily burn when subjected to the heat generated by fires.

These sheets are laid to match metal and fibre cement

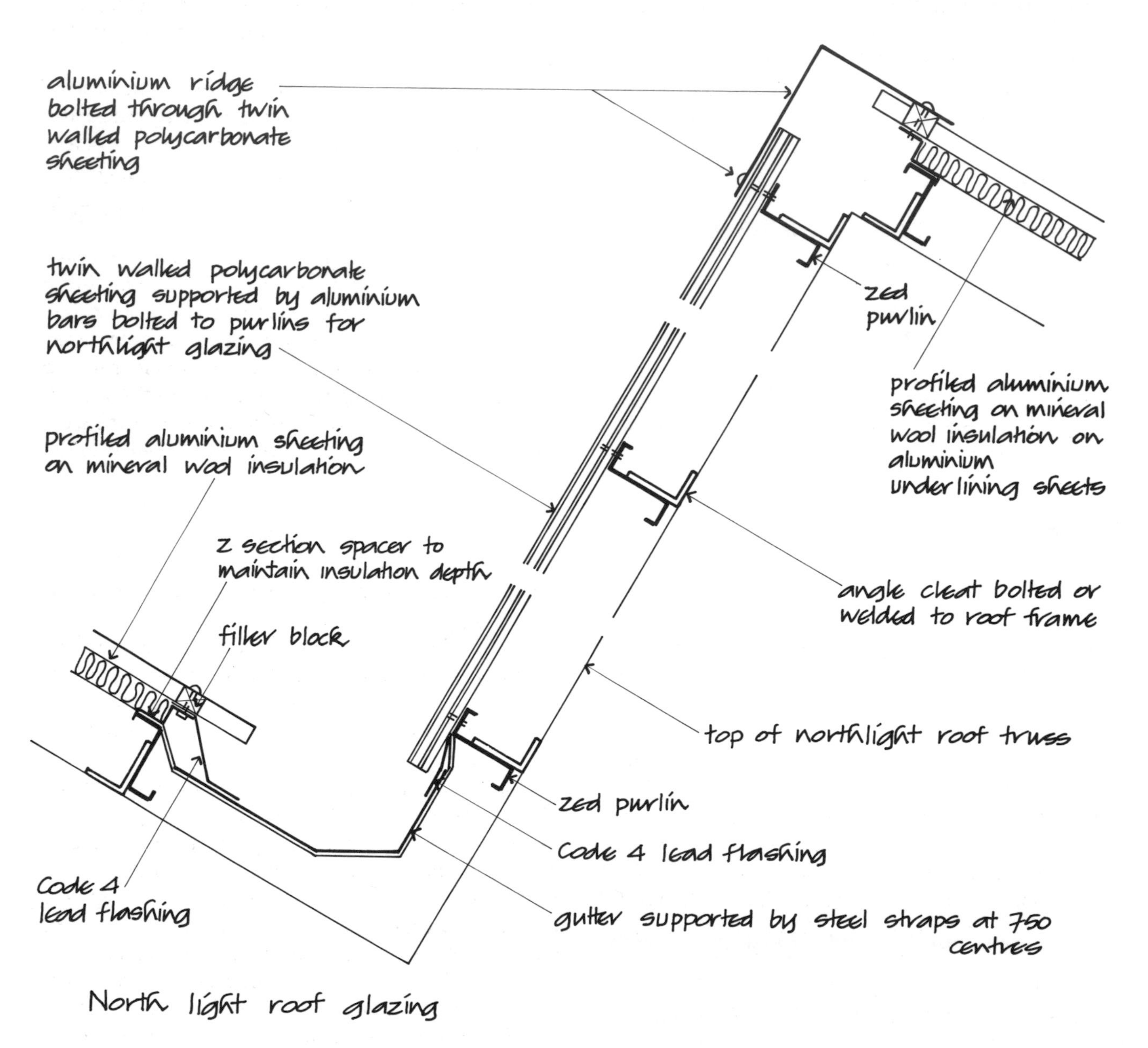

Fig. 95

roofing sheets with side laps of profiles and under and over end laps to match the laps for the abutting roof sheeting. For roof pitches of 15° and less, side and end laps should be sealed with sealing strips and all laps between uPVC sheets should be sealed.

Uplift due to wind suction dictates the necessary centres of fixing for fasteners that should be fitted to holes in the sheets that are 3 larger than the fastener to allow for the considerable thermal expansion of the material. Fasteners similar to those used for roofing sheets are used. Double skin rooflights are formed with two sheets of profiled uPVC with plastic spacers or as sealed double skin rooflights with a profiled top sheet and a flat undersheet. The details shown in Fig. 94 apply both to GRP and uPVC rooflights.

Transparent double or triple skin cellular flat sheets of polycarbonate are used for rooflights because of the extremely good impact resistance of the material. The light transmission of the clear sheet is about 80%. Because of the cellular structure, these flat sheets have good strength and stability and a *U* value for the double skin of 2.8 W/m^{2}K. In common with other plastic materials polycarbonate softens when subjected to the heat generated by fires. Polycarbonate sheeting is more expensive than either uPVC or GRP.

The flat cellular sheets of polycarbonate are supported by aluminium glazing bars fixed to purlins. The capping of the glazing bars compresses a neoprene gasket to the sheets to make a weathertight seal. Figure 95 is an illustration of polycarbonate sheeting used as a rooflight to the north facing slope of a roof.

Patent glazing

The traditional method of fixing glass in the slopes of roofs as rooflights is by means of wood or metal glazing bars that provide support for the glass and form weather flashings or cappings to exclude rain. The word 'patent' refers to the patents taken out by the original makers of glazing bars for rooflights. The original wood, iron and steel glazing bars have been replaced by aluminium and lead or plastic coated steel bars. The disadvantage of patent glazing is the considerable labour and expense in the provision and fixing of glazing bars at comparatively close centres and the necessary top and bottom flashings to weather the overlap with roof sheeting. The advantage of patent glazing is that glass maintains its hard, lustrous, fire-glazed finish which is easy to clean and does not discolour and so reduce light transmission during the useful life of buildings. For this reason glass is sometimes preferred as a glazing material for the roofs and walls of buildings.

Glass has poor resistance to the transfer of heat, the *U* value of single 6 thick glass being 5.7 W/m^{2}K and that of double-glazing 2.8 W/m^{2}K. Glass is a comparatively heavy glazing material being 15 kg/m^2 for 6 thick glass.

When subjected to the heat generated by fires, ordinary glass shatters and falls. Wired glass is used in rooflights because the wire embedded in the glass keeps it in place for some time once the glass shatters in the heat of fires, thus reducing the hazard from falling glass and maintaining the glass as a barrier to the spread of fire.

The most commonly used glazing bars are of extruded aluminium with seatings for glass, condensation channels and a deep web top flange for strength and stiffness in supporting the weight of glass. Glass is secured with clips, beads or screwed or snap-on cappings.

Figure 96 is an illustration of aluminium glazing bars used to support single sheet wired glass as rooflights in the slope of a symmetrical pitch roof. The glazing bars are secured in fixing shoes screwed or bolted to angles fixed to purlins and fitted with aluminium stops to prevent glass slipping down the slope of the roof. Aluminium spring clips, fitted to grooves in the bars, keep the glass in place and serve as weathering between the glass and the bar.

A system of steel battens and angles and an angle and a purlin provide a fixing for glass and sheeting at the overlap of the rooflight and the sheeting as illustrated in Fig. 96. Lead flashings are fixed as weathering at the overlap of glass and sheeting.

Figure 97 is an illustration of an aluminium glazing bar for single-glazing and an aluminium glazing bar for sealed double-glazing units that are secured with aluminium beads bolted to the bar and weathered with butyl strips.

Figures 98 and 99 are illustrations of aluminium glazing bars with bolted aluminium capping and snap-on aluminium to the bars. Cappings are used to secure glass in position on steep slopes and for vertical glazing as they afford a more secure fixing than spring clips and also for appearance to give more emphasis to the bars which would otherwise look somewhat insignificant.

Figures 100 and 101 are illustrations of steel bars covered with lead and PVC sheathing as a protection against corrosion. Steel bars are used for the mechanical strength of the material and the advantage of more

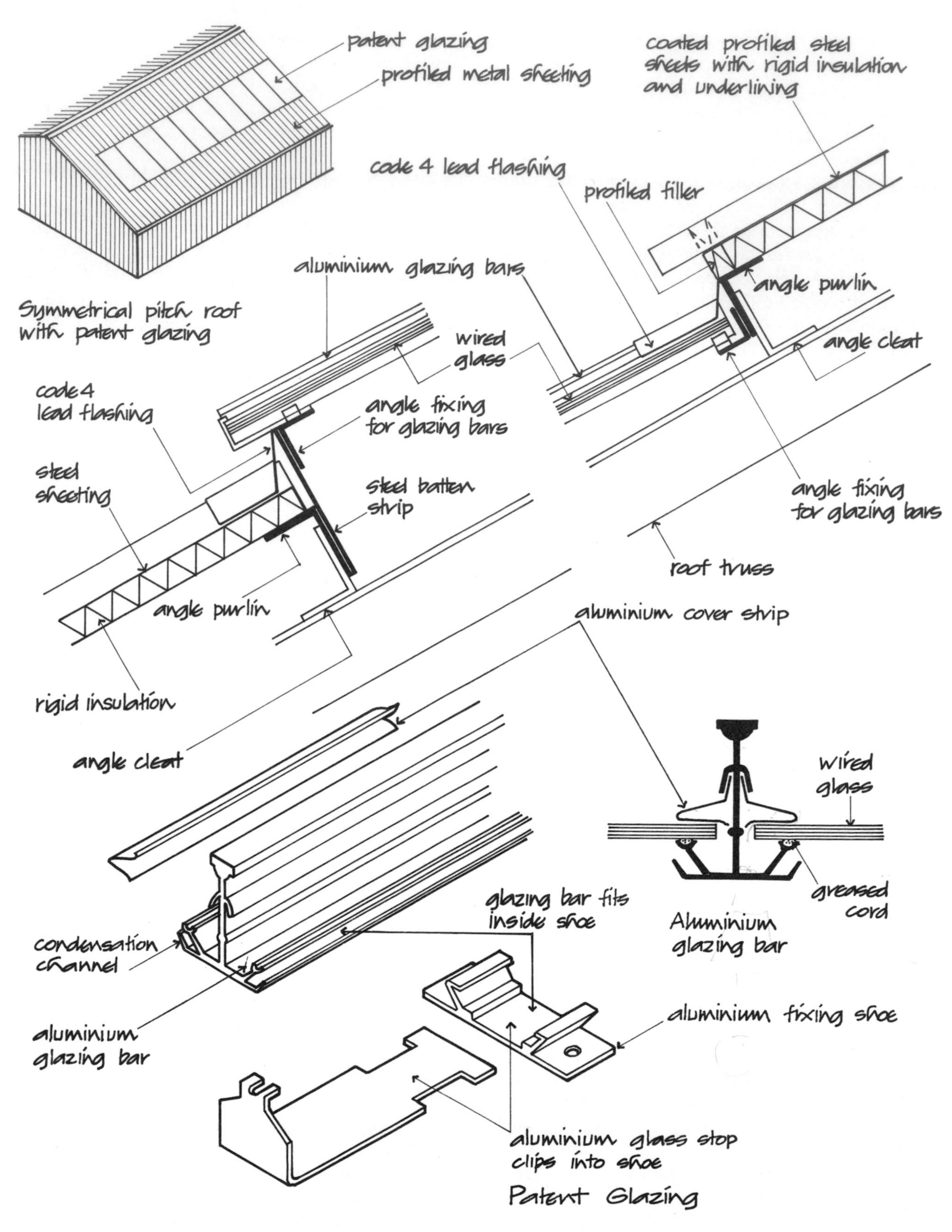

patent glazing
profiled metal sheeting
Symmetrical pitch roof with patent glazing
coated profiled steel sheets with rigid insulation and underlining
code 4 lead flashing
profiled filler
aluminium glazing bars
angle purlin
wired glass
angle cleat
code 4 lead flashing
angle fixing for glazing bars
steel sheeting
steel batten strip
angle fixing for glazing bars
roof truss
angle purlin
aluminium cover strip
rigid insulation
angle cleat
wired glass
greased cord
Aluminium glazing bar
glazing bar fits inside shoe
condensation channel
aluminium fixing shoe
aluminium glazing bar
aluminium glass stop clips into shoe
Patent Glazing

Fig. 96

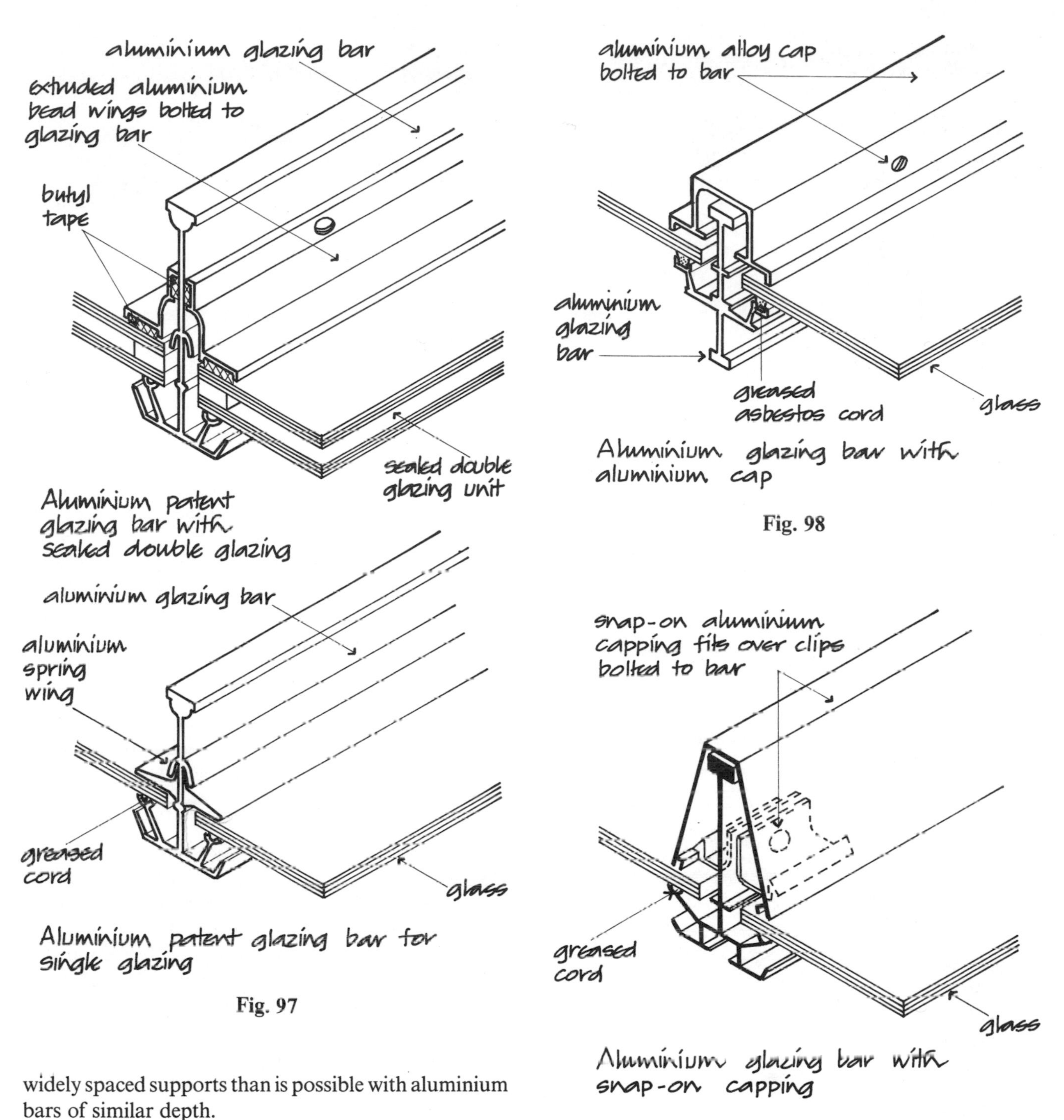

Fig. 97

Fig. 98

Fig. 99

widely spaced supports than is possible with aluminium bars of similar depth.

Rooflights in flat and low pitch roofs

Before the introduction of plastic and fibre glass as material for rooflights, the majority of rooflights to flat roofs were constructed as lantern lights or deck lights which were framed in timber or steel and covered with glass.

A lantern light is constructed with glazed vertical sides and a hipped or gable-ended glazed roof. The vertical sides of the lantern light are used as opening lights for ventilation as illustrated in Fig. 102. Lantern lights were often used to cover considerable areas, the light being framed with substantial timbers of iron or

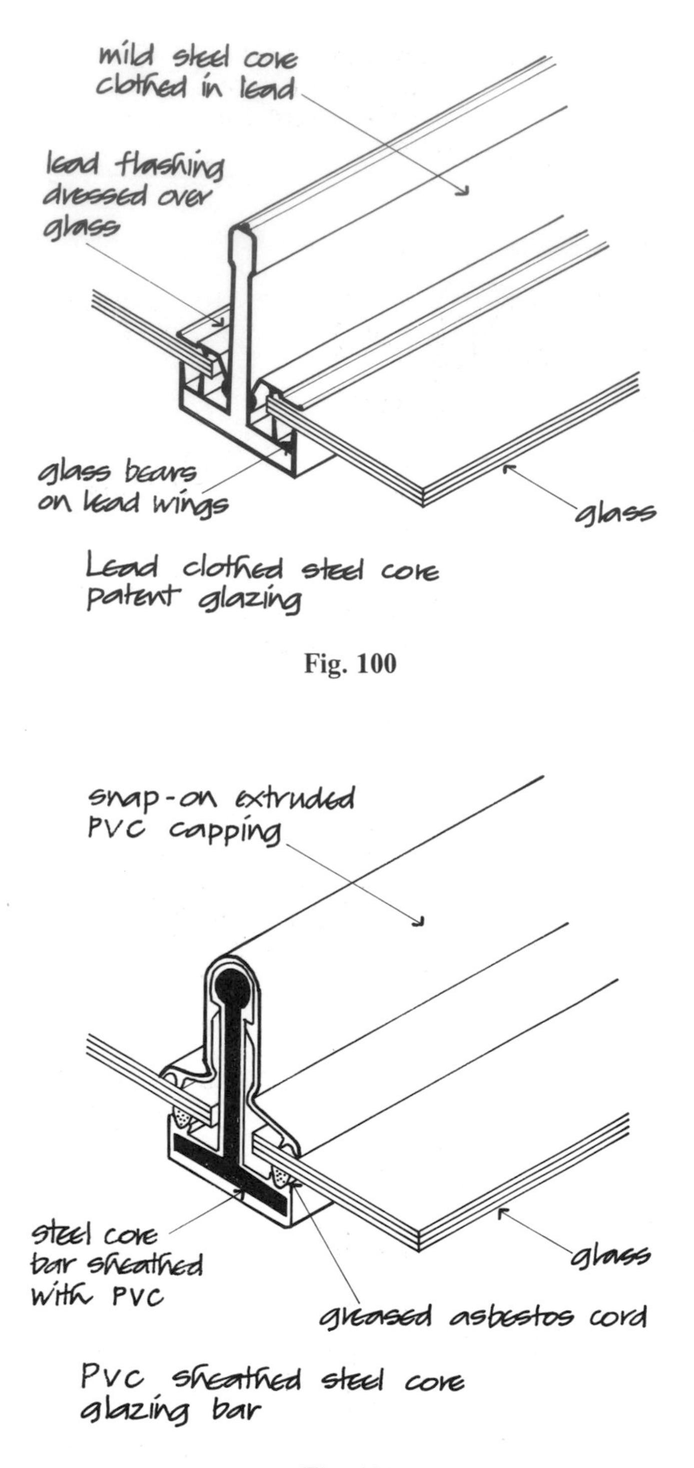

Fig. 100

Fig. 101

steel and frames in the form of a glazed roof to provide top light to large stair-wells and large internal rooms. The traditional lantern light of timber or steel requires frequent and careful maintenance if it is to remain sound and watertight. Lantern lights have largely been

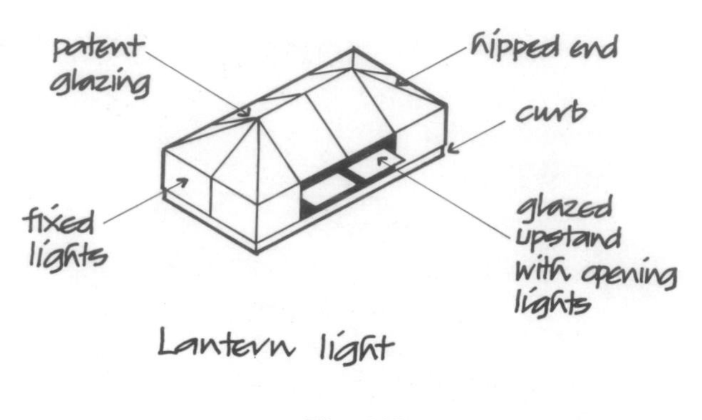

Fig. 102

replaced by domelights for economy in first cost and freedom from maintenance. The advantage of the lantern light is the facility of ventilation from the opening upstand sides that can be controlled by cord or winding gear from below to suit the occupants of the room or space.

Figure 103 is an illustration of an aluminium lantern light constructed with standard aluminium window frame and sash sections, aluminium corner posts and aluminium patent glazing to the pitched roof with an aluminium ridge section. The aluminium sections require no maintenance other than occasional washing. In common with all rooflights fixed in flat roofs, the lantern light illustrated in Fig. 103 is bolted to an upstand curb against wind uplift, to which the upstand skirting of the roof covering is dressed to a height of at least 150.

Deck lights are constructed as a hipped or gable-ended glazed roof with no upstand sides. The deck light does not provide a means of ventilation and serves solely as a rooflight as shown in Fig. 104. The deck light illustrated in Fig. 105 is constructed with lead sheathed steel glazing bars pitched and fixed to a ridge and bolted to a steel tee fixed to the upstand curb. The monopitch light, illustrated in Fig. 106, combines the simplicity of construction of a single slope for roof lighting with the advantage of one glazed upstand side for ventilation from one direction only.

The nature of the materials, glass, reinforced plastic (GRP), uPVC, acrylic and polycarbonate, facilitates the production of a range of shaped rooflights for use in flat and low pitched roofs. The disadvantage of some of the plastic materials is that they discolour and may require replacement after some ten years to restore daylight penetration.

The advantage of the square and rectangular base rooflights, illustrated in Fig. 107, is that they require

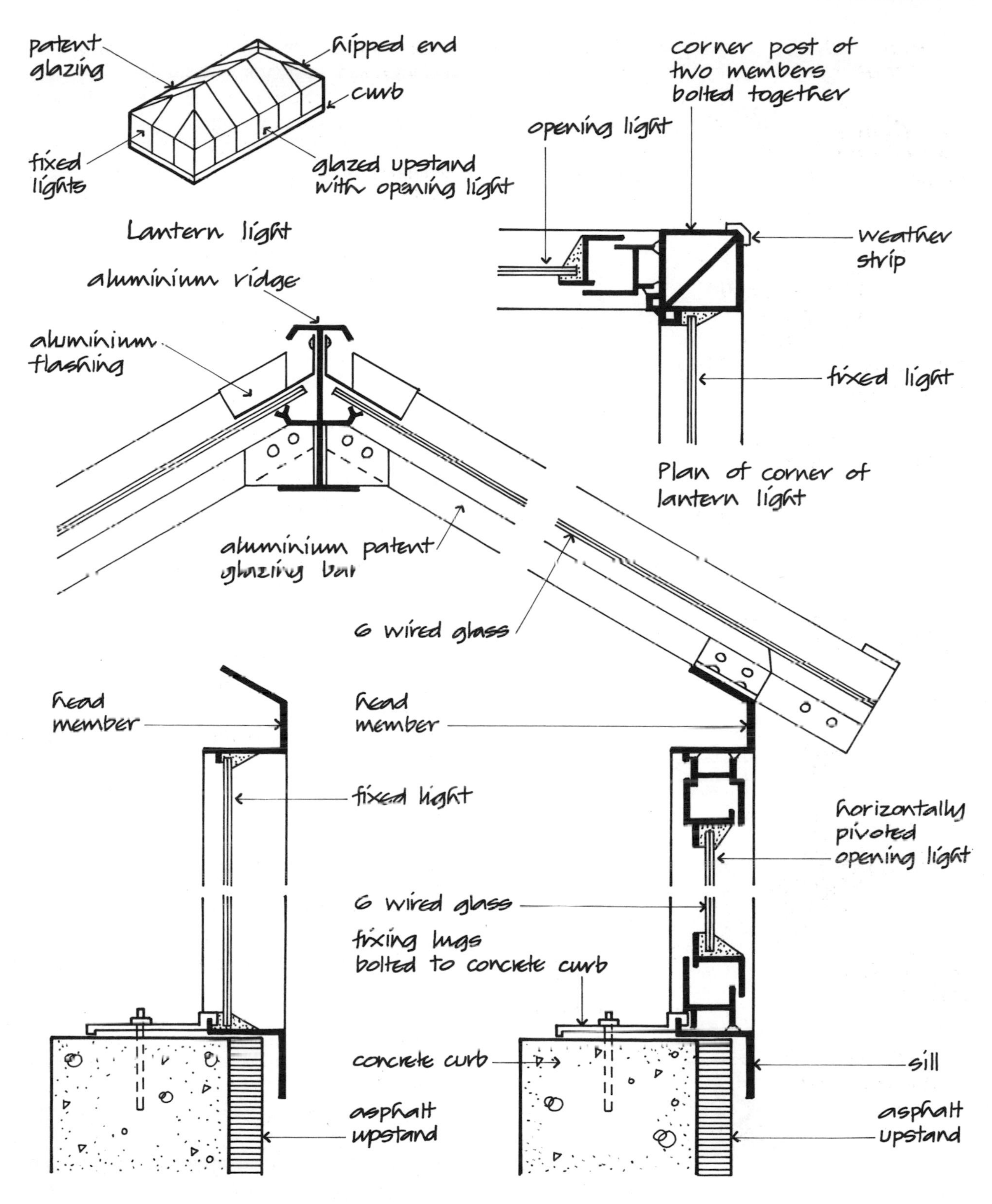
patent glazing
hipped end
curb
fixed lights
glazed upstand with opening light
Lantern light
corner post of two members bolted together
opening light
weather strip
fixed light
Plan of corner of lantern light
aluminium ridge
aluminium flashing
aluminium patent glazing bar
6 wired glass
head member
head member
fixed light
horizontally pivoted opening light
6 wired glass
fixing lugs bolted to concrete curb
concrete curb
sill
asphalt upstand
asphalt upstand

Aluminium lantern light

Fig. 103

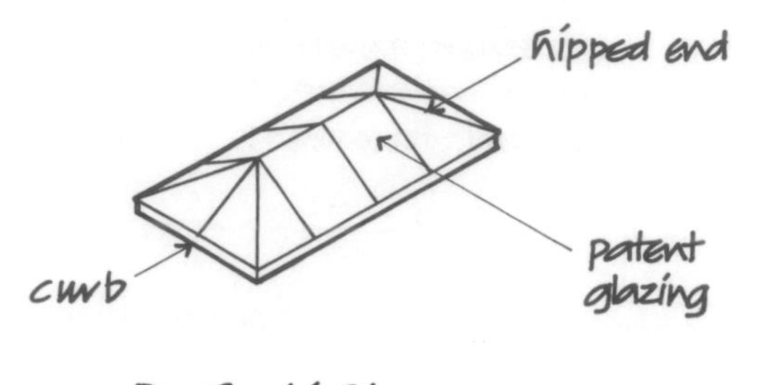

Fig. 104

straightforward trimming of the roof structure around openings and upstands and flashings to curbs as compared to the more complicated trimming and flashings required around the circular base of domelights. One of the commonly used rooflights is the rectangular base domelight illustrated in Fig. 107 which can be formed in one piece or made up in sections and joined with glazing bars to cover larger openings. The advantages of these lights are that they are economical to manufacture and fix, are lightweight and have

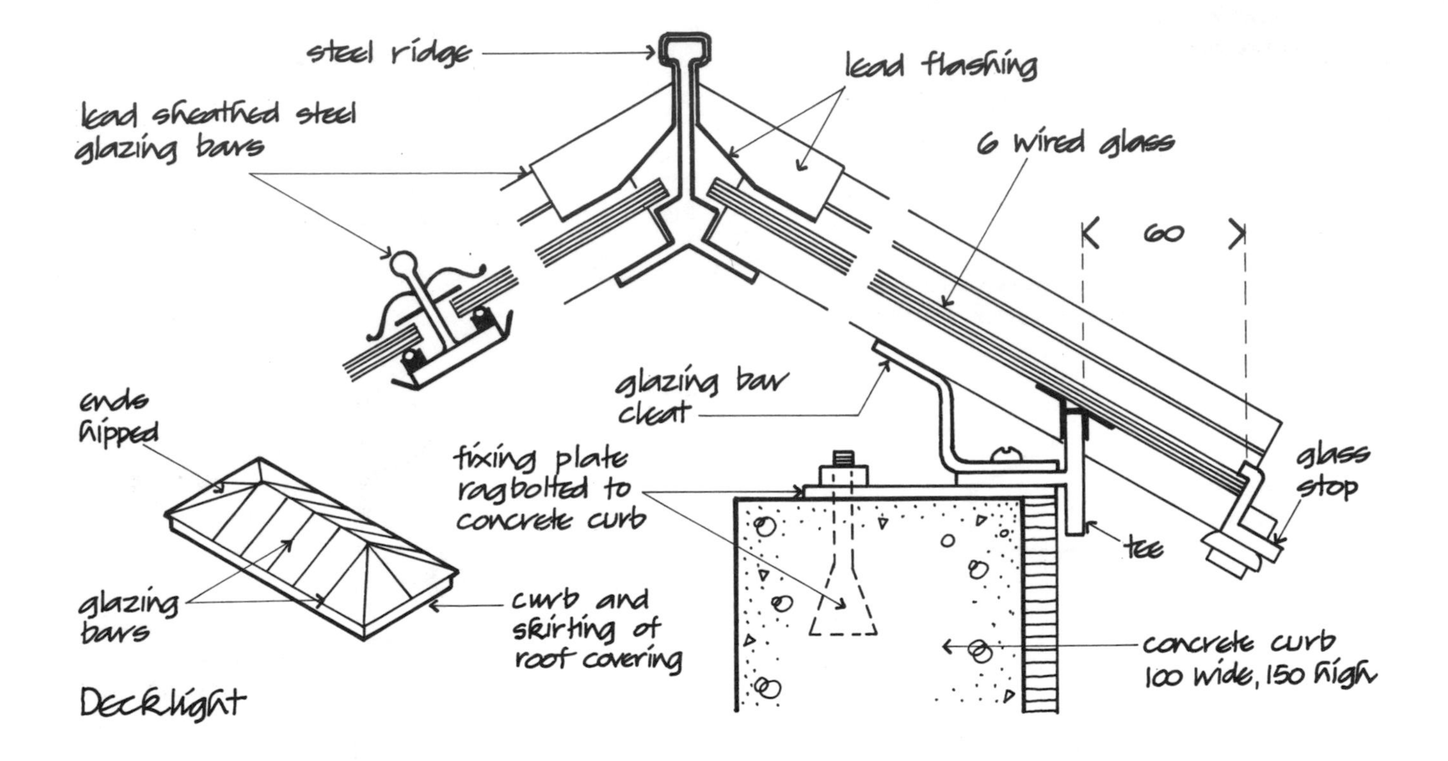

Fig. 105

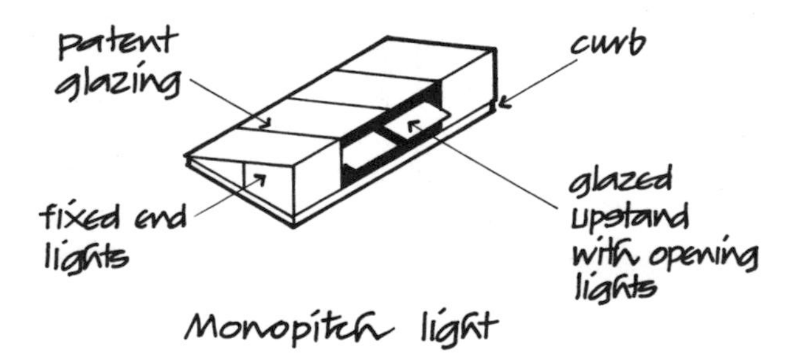

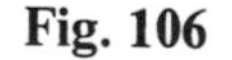
Fig. 106

adequate strength and stiffness from the curved shape of the light.

The round base domelight, shown in Fig. 107, is more expensive to construct than a square base light because of the additional labour involved in trimming a round opening in a roof.

The pyramid light, illustrated in Fig. 107, is used for appearance as the steeply pitched glazed sides afford no increase in light penetration through the opening in the roof.

Plastic rooflights are made as either single skin lights

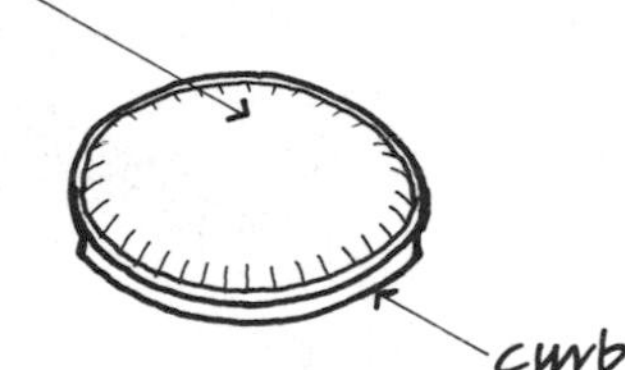

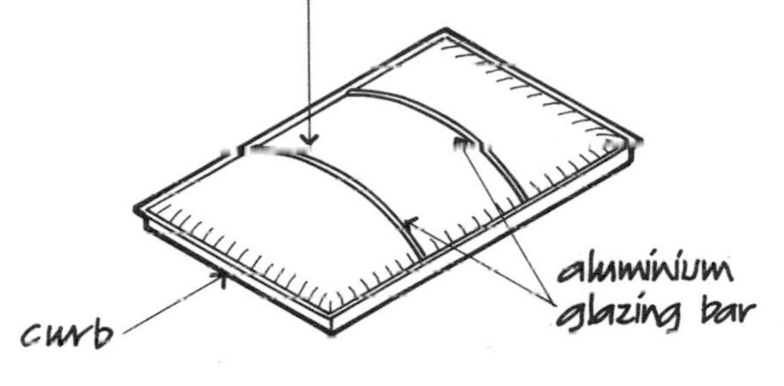

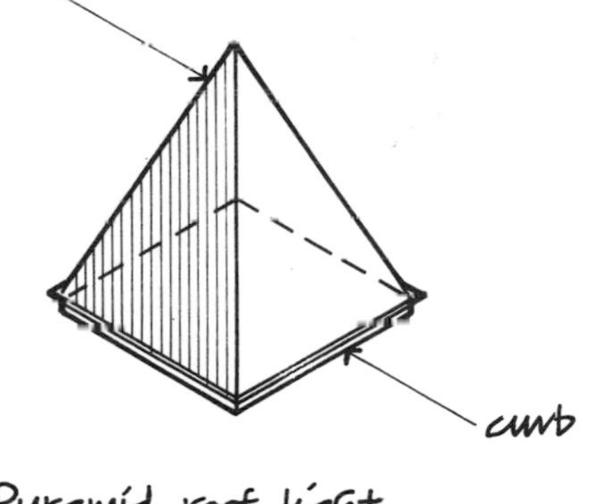

Fig. 107

or as sealed double skin lights which improves their resistance to the transfer of heat. Plastic rooflights are bolted or screwed to upstand curbs against wind uplift, formed on the roof to which an upstand skirting of the roof covering is dressed as illustrated in Fig. 108.

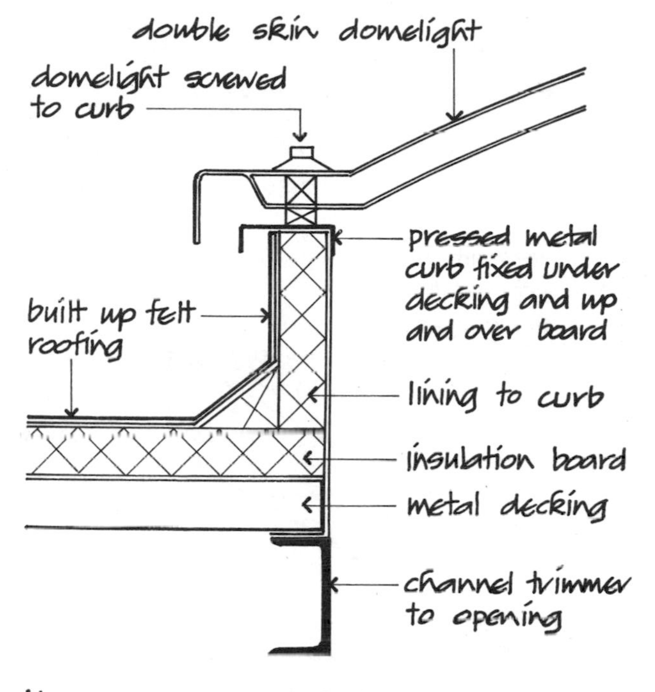

Fig. 108

Lens lights

Lens lights consist of square or round glass blocks or lenses that are cast into reinforced concrete ribs, as illustrated in Fig. 109, to provide diffused daylight through concrete roofs. The lens lights can be pre-cast and bedded in place on site or in-situ cast in a concrete roof. The daylight transmission of these lights is poor compared to other forms of rooflight. Lens lights are used in a concrete roof as rooflights to provide resistance to fire, for reasons of security and to reduce sound transmission.

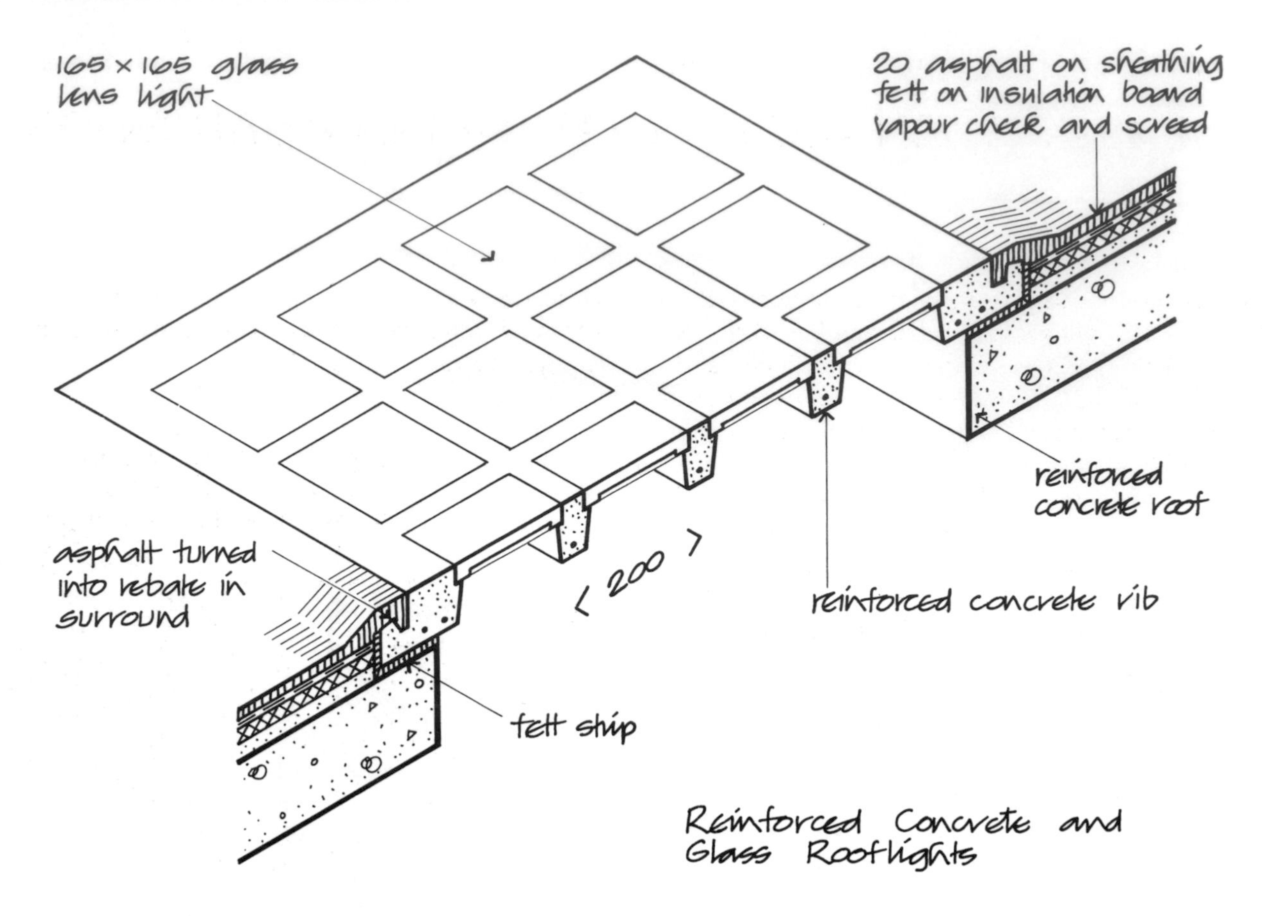
165 × 165 glass lens light
20 asphalt on sheathing felt on insulation board vapour check and screed
reinforced concrete roof
asphalt turned into rebate in surround
200
reinforced concrete rib
felt strip
Reinforced Concrete and Glass Rooflights

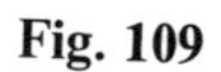
Fig. 109

CHAPTER FOUR

DIAPHRAGM, FIN WALL AND TILT-UP CONSTRUCTION

Brickwork has for centuries been the traditional material for the walls of houses and other small buildings. A one brick thick wall of well burned bricks, bonded and laid in a mortar of the same density and porosity as the bricks, has more than adequate strength to support the comparatively small loads from the floors and roof of a house, and sufficient stability in resisting the lateral pressure of wind. The wall, either solid or more usually as a cavity wall, will resist penetration of rain, have good resistance to damage by fire, require very little maintenance and have a useful life in excess of more than 100 years.

Brickwork has good compressive strength in supporting vertical loads but poor tensile strength in resisting lateral pressure from the lateral loads of floors and roof and wind pressure. The minimum thickness of walls is prescribed in the Building Regulations by reference to the height and length of walls so that the greater the height of wall the greater the thickness the wall has to be at its base, to resist the lateral forces that tend to overturn it (see Volume 1).

A brick wall acts structurally as a vertical cantilever, rising vertically from its fixed base on the foundation so that lateral forces, such as wind, tend to cause the wall to bend. This bending is resisted by the small tensile strength of the brickwork, lateral restraint by floors and roof built into the wall and by buttressing walls and piers built into the wall (see Volume 1). The higher the wall, the greater the vertical cantilever arm of the wall and the thicker the wall needs to be at its base to resist overturning caused by lateral forces.

The majority of tall, single-storey buildings, enclosing large open areas, such as sports halls, warehouses, supermarkets and factories with walls of more than 5.0 in height were until recently built with a frame of lattice steel or a portal frame covered with steel or fibre cement sheeting, insulation and a protective inner lining. Of recent years brick diaphragm or fin walls have been increasingly used for this type of building for the economy, durability, resistance to fire and penetration of rain and thermal and sound insulation advantages of such structures.

A diaphragm wall is built with two leaves of brickwork bonded to brick cross ribs or diaphragms inside a wide cavity between the leaves so that the wall is formed of a series of stiff box or I-sections structurally as illustrated in Fig. 110.

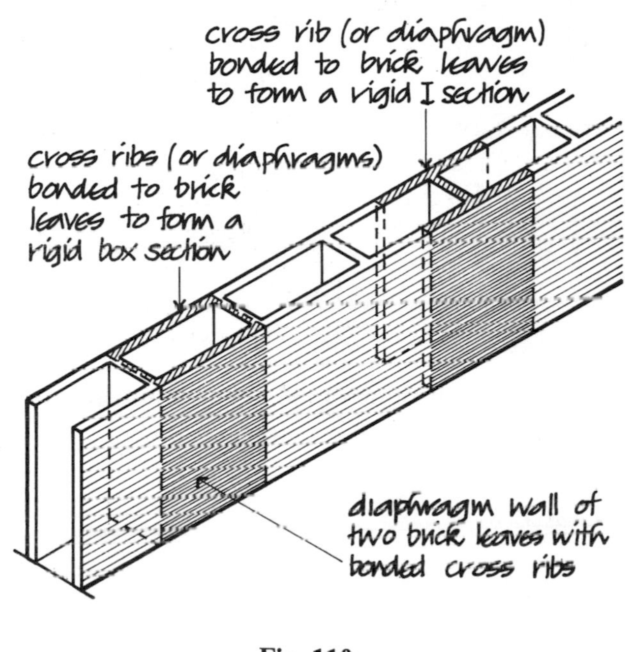

Fig. 110

A fin wall is built as a conventional cavity wall buttressed with piers or fins bonded to the external leaf of the cavity wall to buttress or stiffen the wall against overturning. A fin wall acts structurally as a series of T-sections, as illustrated in Fig. 111. The effective width of the flange of the T-section, that is the outer leaf, may be less than the centres of the fins for design calculations.

BRICK DIAPHRAGM WALLS

The economic advantage of a diaphragm wall, in comparison to a portal frame structure, increases with the height of the wall. For wall heights of up to about 5.0 there is no cost benefit in using a diaphragm wall instead of a portal frame. For wall heights of over 5.0 the diaphragm wall is an economic alternative to a

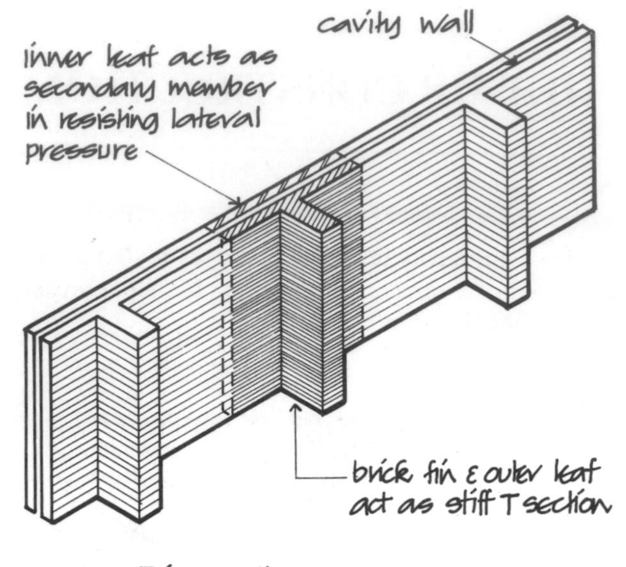

Fig. 111

portal frame structure for tall, single-storey, single-cell buildings.

Strength

The compressive strength of the bricks and mortar of a diaphragm wall is considerable in relation to the comparatively small dead load of the wall, roof and imposed loads of wind and snow.

Stability

A diaphragm wall is designed for stability through the width of the cavity and the spacing of the cross ribs to act as a series of stiff box or I-sections and by the roof which is tied to the top of the wall to act as a horizontal plate to resist lateral forces.

Construction

The width of the cavity and the spacing of the cross ribs is determined by the box or I-section required for stability and the need for economy in the use of materials by using whole bricks whenever possible. Cross ribs are usually spaced four or five whole brick lengths (with mortar joints) apart and the cavity one-and-a-half or two-and-a-half whole bricks (with mortar joints) apart so that the cross ribs can be bonded in alternate courses to the outer and inner leaves. Figure 112 is an illustration of the bonding of typical diaphragm walls. It will be seen that the stretcher bond of the leaves is broken by header faces where the cross ribs are bonded in alternate courses. The colour of the header faces of many bricks is noticeably different from that of the stretcher faces so that in a diaphragm wall where the cross ribs are bonded to the leaves there is a distinct pattern on the wall faces. This pattern can be avoided, for appearance, by bonding the cross ribs to either one or both of the leaves with metal shear ties built into the cross ribs and the leaves.

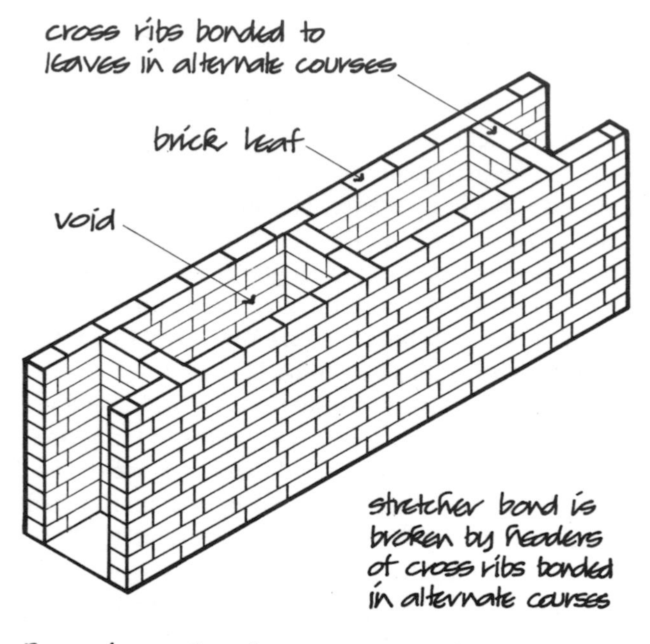

Fig. 112

The loads on the foundation of a diaphragm wall are so slight that a continuous concrete foundation is used for walls built on most natural undisturbed subsoils. A concrete strip foundation to a diaphragm wall is illustrated in Fig. 113. The width of the foundation is determined from the load on the foundation and the safe bearing capacity of the subsoil (see Volume 1).

The roof of a diaphragm wall is tied to the top of the wall to act as a prop in resisting the overturning action

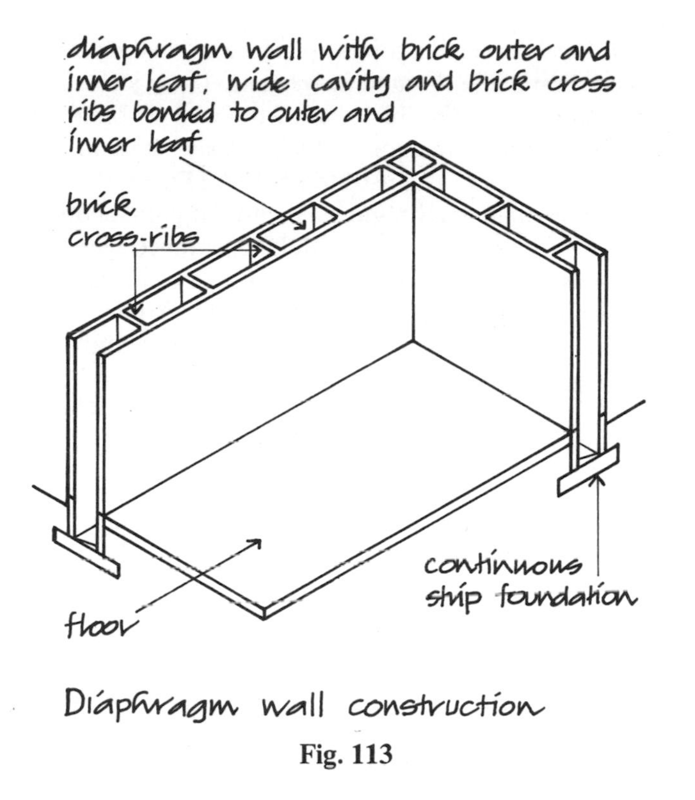

Fig. 113

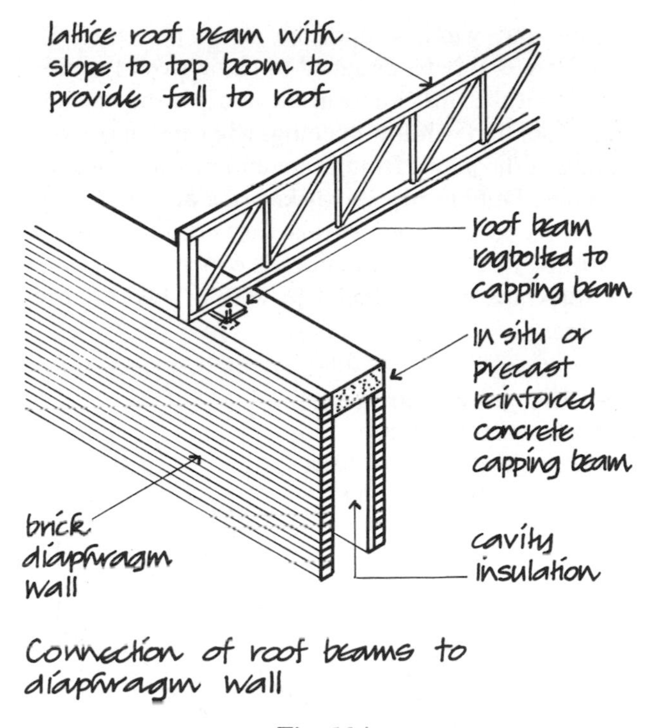

Fig. 114

of lateral wind pressure, by transferring the horizontal forces on the long walls to the end walls of the building that act as shear walls. To ensure that the roof structure is tied to the whole of the length of the top of walls, a reinforced concrete capping beam is cast or bedded on the top of the wall and the roof beams are bolted to the capping beam as illustrated in Fig. 114. It will be seen from Fig. 114 that the reinforced concrete capping does not project to the external face of the wall. This is solely for the sake of appearance.

The capping beam can be of reinforced, in-situ cast, concrete on a support of fibre cement sheet. The disadvantage of this method of construction is the near impossibility of preventing wet cement stains disfiguring fairface brick surfaces below. To avoid unsightly stains on brickwork a system of pre-cast reinforced concrete capping beams is used. The beam is cast in lengths suited to the convenience of transport and handling and to span between cross ribs. The sections are tied with end ties or anchor bolts cast into the ends of sections and the joint is made with cement grout. Roof beams are tied to the capping beam with studs cast in the beam to which the beam is bolted as illustrated in Fig. 114.

The roof is tied to the capping beam to act as a horizontal prop to the top of the wall by transferring loads to the end walls. So that roof beams act together as a stiff plate they are braced by horizontal lattice steel wind girders connected to the roof beams, as illustrated in Fig. 115.

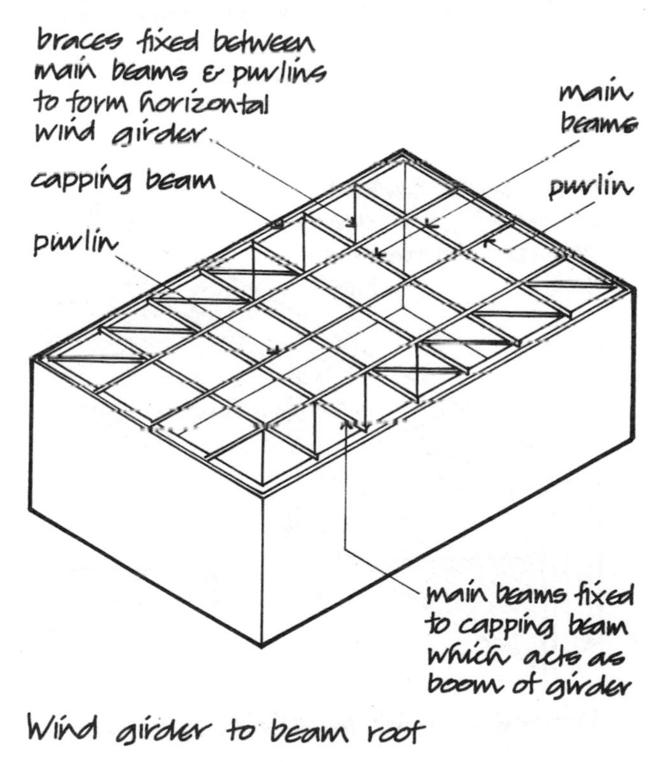

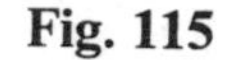

Fig. 115

Door and window openings in diaphragm walls should preferably be designed to fit between the cross ribs so that the ribs can form the jamb of the opening. Large door and window openings will cause large local loading at the jambs from beam end bearings over the openings. Double ribs or thicker ribs are built at the jambs of large openings, to take the additional load, as illustrated in Fig. 116. Reinforced concrete lintels, the full thickness of the wall, are cast or bedded over openings.

Vertical movement joints are necessary in long walls to accommodate thermal movement. These joints are formed through the thickness of the wall by building double ribs to form a joint which is sealed with a non-hardening mastic as illustrated in Fig. 117.

A long, high diaphragm wall with flat parallel brick leaves may have a somewhat dull appearance. The flat surfaces of the wall can be broken by the use of projecting brick piers and a brick plinth, as illustrated in Fig. 118, where selected cross ribs project from the

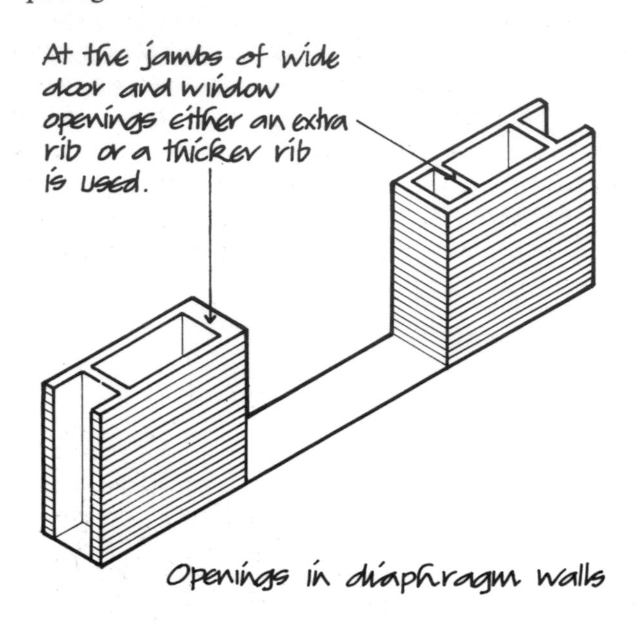

Fig. 116

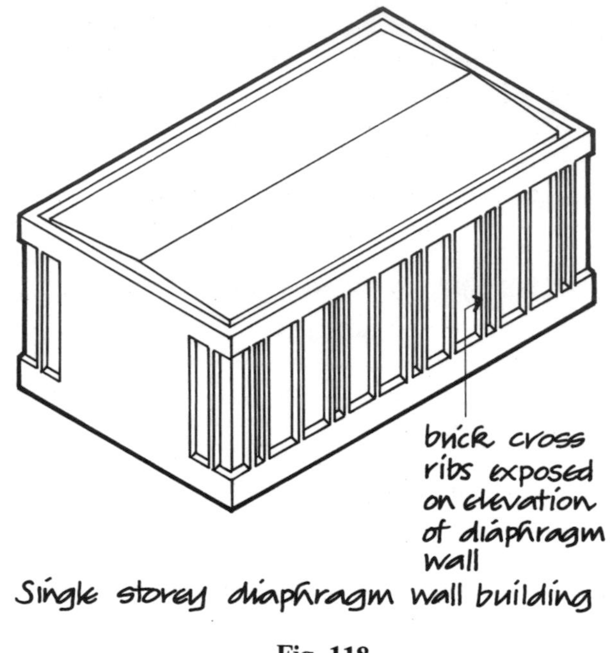

Fig. 118

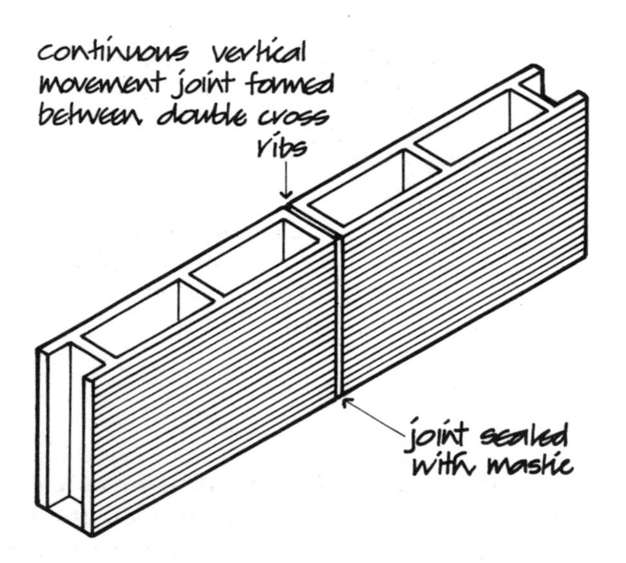

Fig. 117

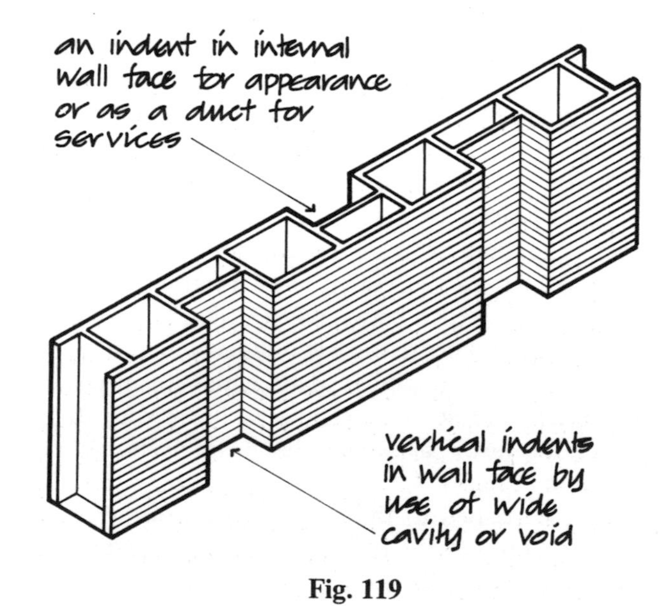

Fig. 119

face of the wall, purely for appearance. As an alternative the external face of the wall can be indented by variations in the width of the cavity as illustrated in Fig. 119. The width and the depth of the breaks in the wall face are chosen for appearance.

Resistance to weather

Experience of diaphragm walls built in positions of severe exposure, and recent tests carried out, show that a diaphragm wall will satisfactorily resist penetration of rainwater to the inner face of the wall. In positions of severe exposure there may be some penetration of rain into the brick cross ribs. To assist drying out of the cross ribs, by evaporation, it may be wise to ventilate the voids in the wall.

A continuous horizontal damp proof course (d.p.c.) should be built into a diaphragm wall for both leaves and cross ribs at floor level and at least 150 above ground level. Bitumen felt d.p.c.s, which have poor resistance to compression, may squeeze out under heavy loads, whereas the more expensive brick d.p.c. of three courses of engineering bricks will provide good resistance to tensile stress at the base of the wall.

At the jambs of openings it is practice to build in a vertical d.p.c. of bitumen felt.

Resistance to the passage of heat

The thermal insulation of a diaphragm wall is some 10% less than that of a conventional cavity wall due to the circulation of air in the voids in the wall. The common method of improving the thermal insulation of a diaphragm wall is by fixing insulating boards, 75 or 100 thick, inside the voids against the inner leaf. The insulating boards are secured in position behind wall ties built into the cross ribs or with galvanised nails driven into the inner leaf.

The roof of diaphragm wall structures should act as a horizontal plate to prop the top of the wall against lateral forces. Some form of flat roof construction of main beams with or without secondary beams is most suited to act as a plate. Solid-web castellated or lattice main beams spanning the least width of the building, with horizontal wind girders, is the usual roof construction, with metal decking, insulation and built-up bitumen felt roof covering. Laminated timber main beams are used for the appearance of the natural material of the beams which are exposed.

Where pitched roof construction is used the frames of the roof structure should be braced to act as horizontal or near horizontal wind girders to prop the walls.

BRICK FIN WALLS

A fin wall is a conventional cavity wall buttressed by brick fins bonded to the outer leaf and projecting from the external face of the wall to stiffen high walls against horizontal pressures. Fig. 120 is an illustration of part of a fin wall. The minimum dimensions and spacing of the fins are determined by the cross-sectional area of the T-section of the wall required to resist the tensile stress from lateral pressure and by considerations of the appearance of the building. The spacing and dimensions of the fins can be varied to suit a chosen external appearance.

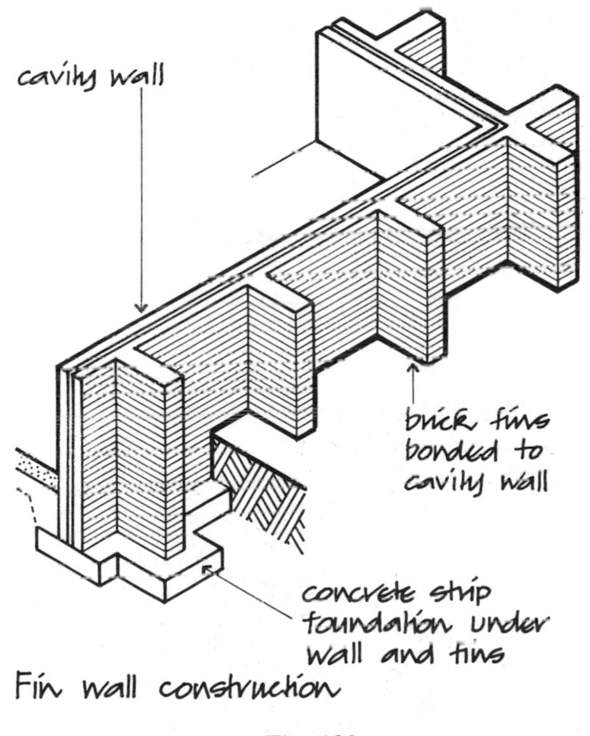

Fig. 120

For walls over about 5.0, a fin wall is used instead of a diaphragm wall because of the effect of the protruding vertical fins illustrated in Fig. 121 which can be built in a variety of ways for appearance. Some typical profiles for brick fins are illustrated in Fig. 122. For best effect, special bricks are used. These special bricks can

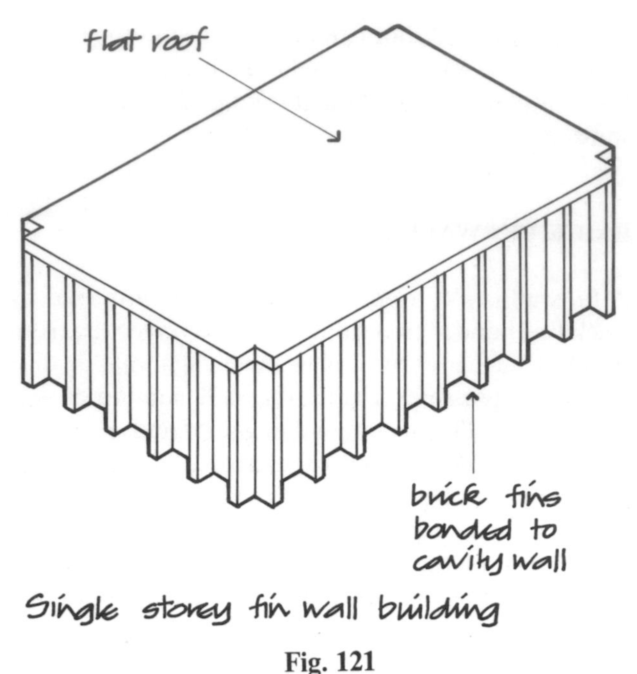

Fig. 121

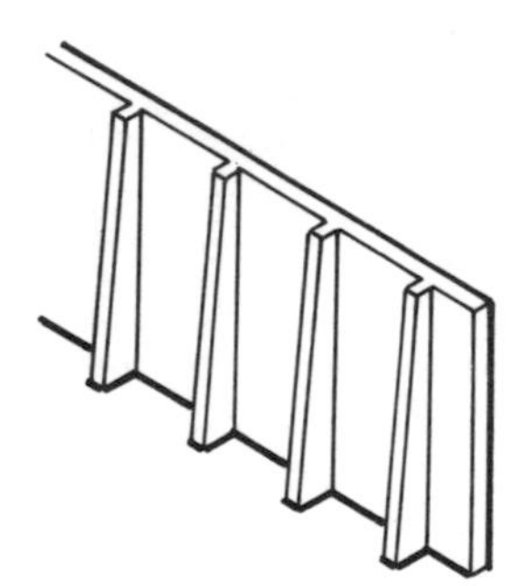

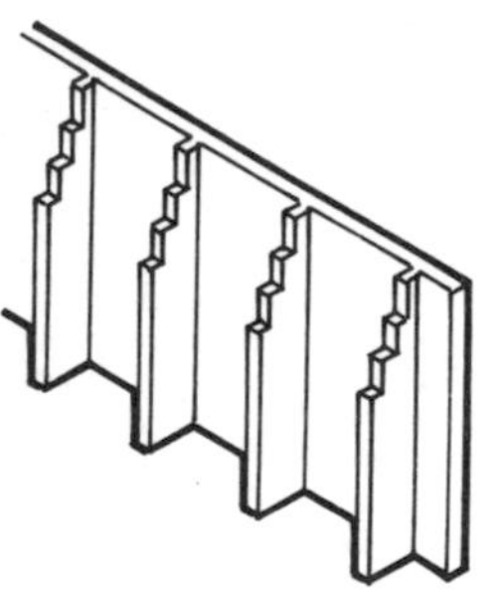

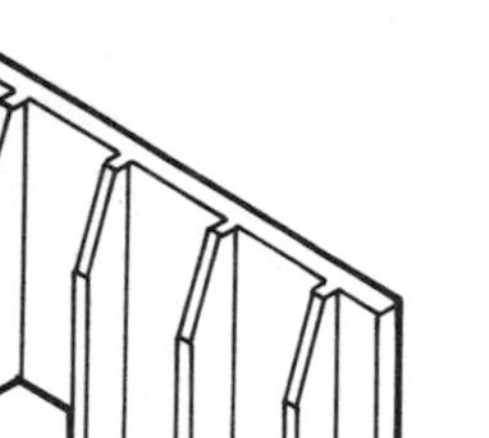

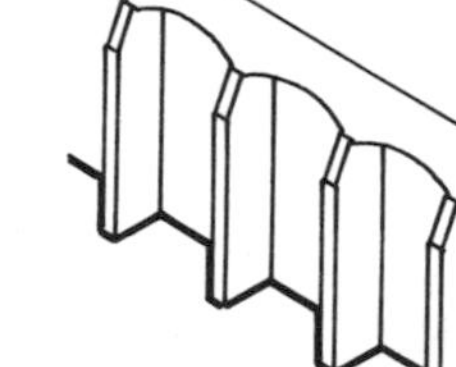

Fig. 122

be selected from the range of 'specials' produced by brickmakers or they can be specially made to order. The use of 'specials' will provide a better finish to brickwork than is possible by cutting standard bricks to the required shape. The use of special bricks does considerably increase the cost of brickwork.

Strength

The strength of the brickwork of a fin wall is considerable in relation to the comparatively small dead load of the wall and roof and imposed loads of wind and snow.

Stability

Stability against lateral forces from wind pressure is provided by the T-section of the fins and the prop effect of the roof, which is usually tied to the top of the wall to act as a horizontal plate to transfer moments to the end walls.

Construction

The wall is constructed as a conventional cavity wall with a 50 cavity and inner and outer leaves of brick tied with wall ties. The fins, which are bonded to the outer leaf in alternate courses, are usually one brick thick

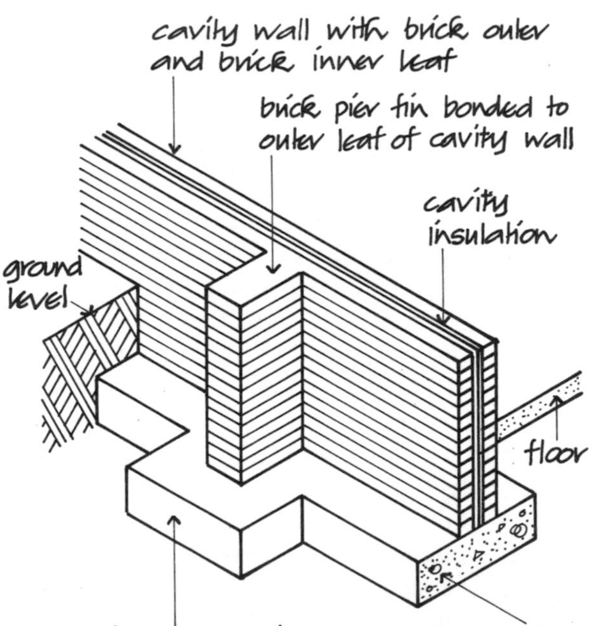

Fig. 123

with a projection of four or more brick lengths. The fins should be spaced a number of whole bricks apart to minimise cutting of bricks and at centres necessary for stability and for appearance.

The loads on the foundation of a fin wall are so slight that a continuous concrete strip foundation will provide support and stability for the wall on most natural subsoils. A continuous strip foundation to a fin wall is illustrated in Fig. 123 from which it will be seen that the foundation is spread under the wall and extended under the fins.

The roof of a fin wall is usually designed as a horizontal plate which props the top of the wall to transfer lateral pressures and so achieve an economy in the required wall section. Roof beams generally coincide with the centres of the fins, the roof beams being tied either to a continuous reinforced concrete capping beam cast or bedded on the wall or to concrete padstones cast or bedded on the fins as illustrated in

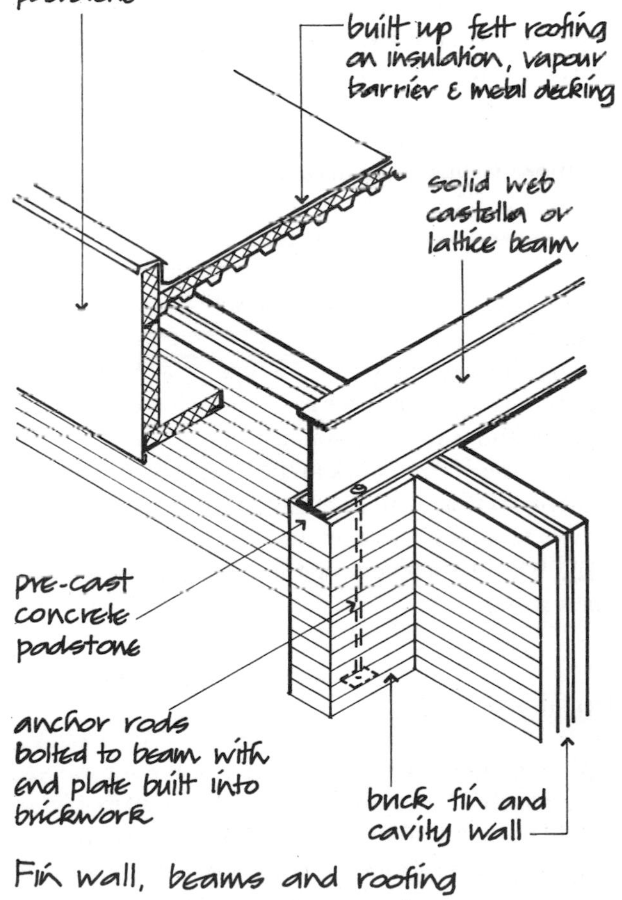

Fig. 124

Fig. 124. To resist wind uplift on lightweight roofs the beams are anchored to the brick fins through bolts built into the fins, cast or threaded through the padstones and bolted to the beams (Fig. 124).

Horizontal bracing to the roof beams is provided by lattice wind girders fixed to the beams to act as a plate in propping the top of the wall. These wind girders are usually combined with a capping plate to the top of fin walls.

Door and window openings in fin walls should be the same as the width between fins for simplicity of construction. To allow sufficient cross-section of brickwork at the jambs of wide openings either a thicker fin or a double fin is built, as illustrated in Fig. 125.

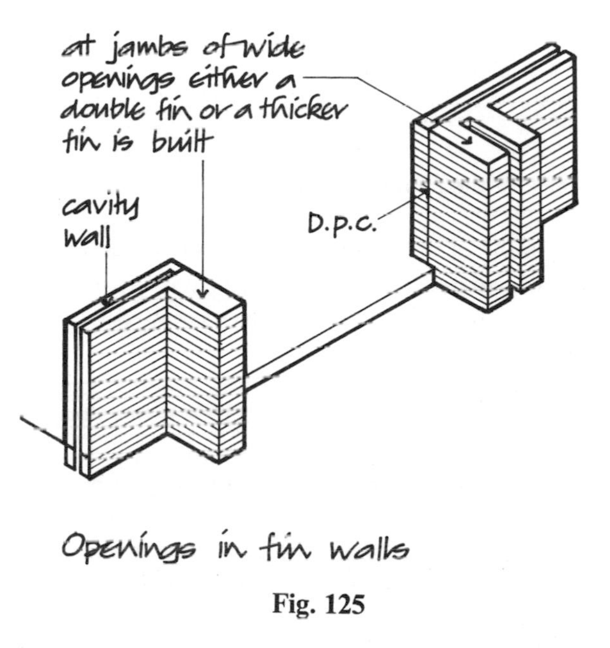

Fig. 125

Movement joints, which are necessary in long walls, are formed between double brick fins as illustrated in Fig. 126.

Resistance to weather

The cavity wall serves as a barrier to the penetration of rain to the inside face of the wall in all but positions of severe exposure. To an extent the projecting fins serve to disperse driving rain and thus give some protection to the cavity wall.

A continuous horizontal damp-proof course must be built into the wall and the fins at floor level and at least 150 above ground. The considerations of the choice of materials are the same as that for diaphragm walls.

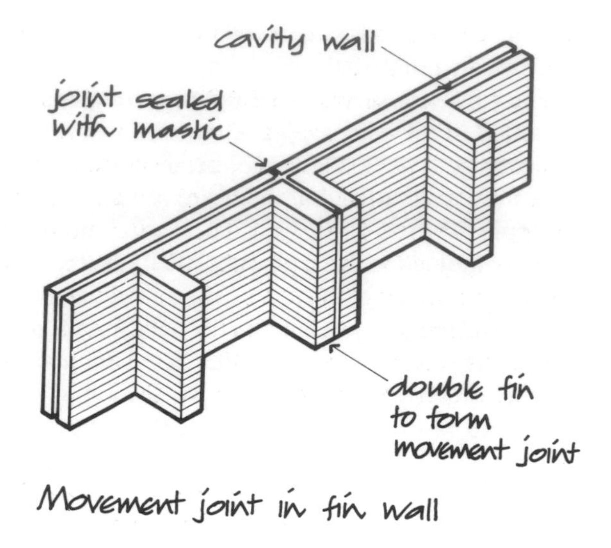

Fig. 126

Resistance to the passage of heat

The insulation of the cavity wall is usually improved by the use of insulation bats fixed in the cavity against the inner leaf of the wall or by the use of cavity fill.

BLOCKWORK WALLS

The majority of diaphragm and fin walls are built of brick because the small unit of the brick facilitates bonding and the construction of fins, recesses and cross ribs. Both diaphragm and fin walls can be built of concrete blocks, where the spacing of cross ribs and fins and width of voids is adapted to the block dimensions to minimise wasteful cutting of blocks. Because of the larger size of blocks, a wall of blocks will tend to have a more massive appearance than a similar wall of bricks.

TILT-UP CONSTRUCTION

Introduction

The term tilt-up construction is used to describe a technique of precasting large, slender reinforced concrete wall panels on site on the floor slab or on a temporary casting bed which, when cured (hardened), are tilted by crane into position as the enclosing wall envelope and structure. This technique or system of building has been used principally for the construction of single-storey commercial and industrial buildings on open sites where there is room for casting and the necessary lifting equipment.

This system of building, which has been much used in the US and Australia for the speed of casting and erecting the panels in a matter of days and for the security, durability and freedom from maintenance of the concrete walls, has more recently been adopted for the construction of multi-storey buildings.

To gain the maximum advantage of speed of casting and erecting that is possible with this system of construction, the individual reinforced concrete wall panels should be cast on the accurately levelled floor slab as close as is practical to their final position.

The site slab of concrete is cast over the completed foundations, drainage and service pipe work and accurately levelled with a wide, travelling screeding machine to provide a level surface onto which the wall panels can be cast. The panels are then cast around reinforcement inside steel edge shuttering on a bond breaker applied to the surface of the site slab. The wall panels may be cast as individual panels, as a continuous strip which is cut to panel size or stack cast one on the other, separated by a bond breaker. After a few days the cured, hardened panels are then lifted into position and propped ready for the roof deck. The sequence of operations is illustrated diagrammatically in Fig. 127.

Panel size is limited by the strength of the reinforced concrete panel necessary to suffer the stresses induced in the panel as it is lifted into the vertical and by the lifting capacity of the cranes. The wall panels may vary in design from plain, flat slabs to frames with wide openings for glazing providing there is adequate reinforced concrete to carry the anticipated loads. A variety of shapes and features is practical and economic by the repetitive use of formwork in the casting bed. A variety of external finishes can be produced from smooth, hard surfaces to a number of textured finishes.

The typical thickness of these reinforced concrete panels is 160.

Strength and stability

Depending on the bearing capacity of the subsoil and the anticipated loads on the foundations, strip, pad or pile and beam foundations may be used. The panels are tilted up and positioned on the levelled foundations against a rebate in the concrete, up to timber runners or onto a seating angle and set level on shims. A

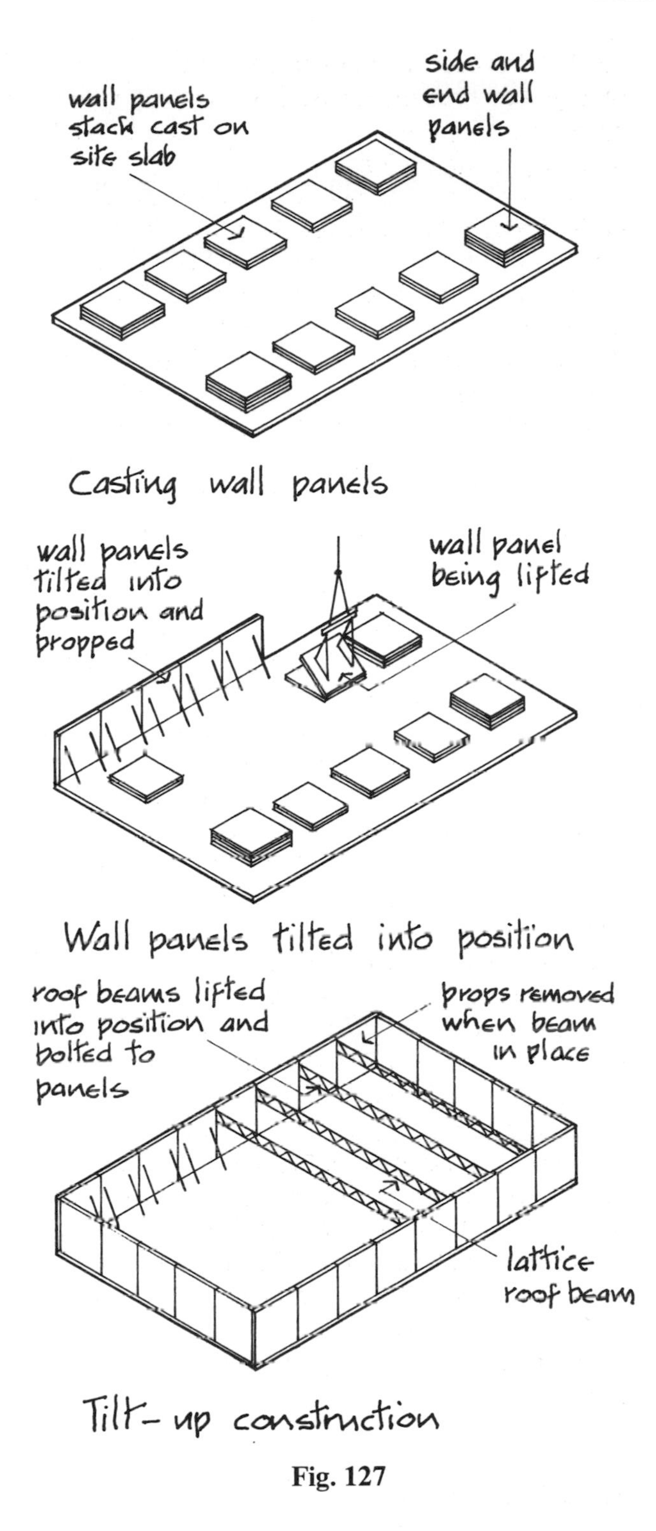

Fig. 127

mechanical connection of the foot of the panels to the foundation or the floor slab is often used. Cast in metal, dowels projecting from the foot of the panels are set into slots or holes in the foundations and grouted in position, or a plate, welded to studs, or bar anchors, cast into the foot of the panel, provide a means of welded connection to rods cast into the site slab as illustrated in Fig. 128.

The roof deck serves as a diaphragm to give support to the top of the wall panels and to transmit lateral wind forces back to the foundation. Lattice beam roof decks are welded to seat angles, welded to a plate and cast in studs as illustrated in Fig. 128. A continuous chord angle is welded to the top of the lattice beams and to bolts cast or fixed in the panel. The chord angle, which serves as a transverse tie across the panels, is secured to the panels with bolts set into slots in the angle to allow for shrinkage movements of the panels.

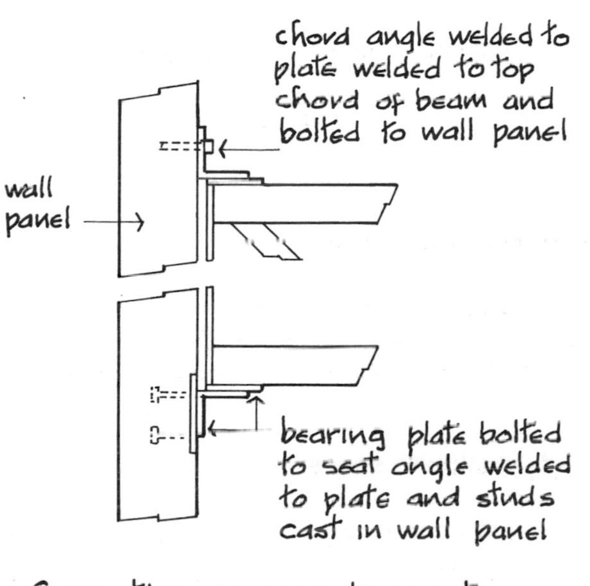

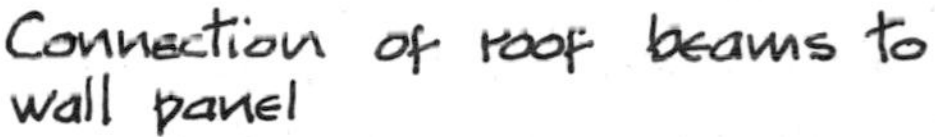

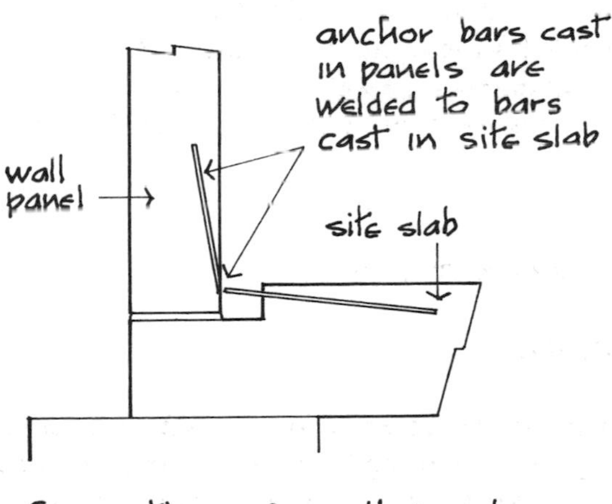

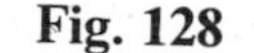
Fig. 128

Resistance to weather

The high density of the concrete panels is adequate to resist the penetration of rainwater to the inside face of the panels. The open vertical joint between panels, which will act as a drainage channel for rainwater falling on the outside face, is sealed against rain and wind penetration with a polysulphide sealant run inside the joint over a polyethylene backing strip.

Durability and freedom from maintenance

A smooth, dense finish to concrete panels should require little if any work of maintenance for many years. This type of finish can be cleaned by washing from time to time or overpainted to enhance appearance.

Fire safety

The reinforced concrete wall panels provide good resistance to damage by fire and will serve as an effective barrier to the spread of fire between buildings.

Resistance to the passage of heat

The usual 160 thickness of the reinforced concrete panels is not sufficient, by itself, to provide the *U* value for walls required by the Building Regulations for buildings which require internal heating. The most straightforward way of enhancing the thermal resistance of the panels is to apply one of the proprietary lining systems to the inner face of the wall to provide an insulating lining and an interior wall finish. Lining panels with a core of foamed PIR or PUR to a plasterboard facing and a polythene moisture vapour check on the inner face are fixed by shot-fired connectors direct to the panels or fixed to timber battens which are shot-fire fixed to the panel.

Resistance to the passage of sound

Solid concrete wall panels without openings provide good resistance to airborne sound either from outside or from inside to outside.

Security

Solid wall panels provide good security against forced entry, damage by vandalism and substantial protection against damage by accidental knocks.

CHAPTER FIVE

SHELL STRUCTURES

For some years from the middle of the twentieth century it was fashionable to construct single-storey buildings as reinforced concrete shells. The majority of shell structures were built in the warm, dry climates of South America and the Middle East where cheap labour made the construction of this form of building reasonably economic. Those buildings that were constructed as shell forms in colder, wetter northern climates were designed as shell forms principally for the sake of the elegant and unusual form that such buildings can take rather than considerations of economy or utility as a weathershield. One of the latest and probably the best known form of shell structure is the Sydney Opera House in Australia.

One of the factors that led to the use of reinforced concrete shell forms was the worldwide shortage of steel that followed the end of World War 2 in the middle of the twentieth century. Reinforced concrete became the principal structural material for framed structures for some years as a substitute for the previously used steel frame. It was a logical development and use of the initially plastic nature of concrete to construct thin shells combining the compressive strength of concrete with the minimum of steel in the form of a shell.

With the current worldwide surplus of steel it is no longer either necessary or economic to use reinforced concrete as a structural material.

A shell structure is a thin, curved membrane or slab, usually of reinforced concrete, that functions both as structure and covering, the structure deriving its strength and rigidity from the curved shell form. The term 'shell' is used to describe these structures by reference to the very considerable strength and rigidity of thin, natural, curved forms such as the shell of an egg, a nut and crustaceans such as the tortoise. The strength and rigidity of curved shell structures makes it possible to construct single curved barrel vaults 60 thick and double curved hyperbolic paraboloids 40 thick in reinforced concrete for spans of 30.0.

The attraction of shell structures lies in the elegant simplicity of curved shell forms that utilise the natural strength and stiffness of shell forms with great economy in the use of material. The disadvantage of shell structures is their cost. A shell structure is more expensive than a portal framed structure covering the same floor area because of the considerable labour required to construct the centering on which the shell is cast.

The material most suited to the construction of a shell structure is concrete, which is a highly plastic material when first mixed with water that can take up any shape on centering or inside formwork. Small section reinforcing bars can readily be bent to follow the curvature of shells. Once the cement has set and the concrete hardened the reinforced concrete membrane or slab acts as a strong, rigid shell which serves as both structure and covering to the building.

Shell structures are sometime described as single or double curvature shells. Single curvature shells, curved on one linear axis, are part of a cylinder or cone in the form of barrel vaults and conoid shells, as illustrated in Fig. 129. Double curvature shells are either part of a sphere, as a dome, or a hyperboloid of revolution, as illustrated in Fig. 129. The terms 'single curvature' and 'double curvature' do not provide a precise geometric distinction between the form of shell structures as a barrel vault is a single curvature shell but so is a dome. These terms are used to differentiate the comparative rigidity of the two forms and the complexity of the centering necessary to construct the shell form. Double curvature of a shell adds considerably to its stiffness, resistance to deformation under load and reduction in the need for restraint against deformation.

Centering is the term used to describe the necessary temporary support on which a curved reinforced concrete shell structure is cast. The centering for a single curvature barrel vault is less complex than that for a dome which is curved from a centre point.

The most straightforward shell construction is the barrel vault, which is part of a cylinder or barrel with the same curvature along its length, as shown in Fig. 130. The short-span barrel vault, illustrated in Fig. 130, is used for the width of the arch ribs between which the barrel vaults span. It is cast on similar arch ribs supporting straight timber or metal centering which is comparatively simple and economic to erect and which can, without waste, be taken down and used again for

Conoid Shell roof
L somewhat less than half S
H about sixth & h ninth of S

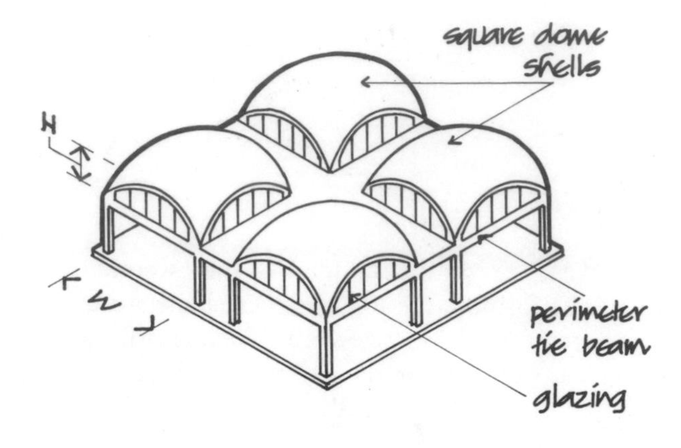

Square Dome Shell roof
radius of domes about six fifths of W
H one tenth of W

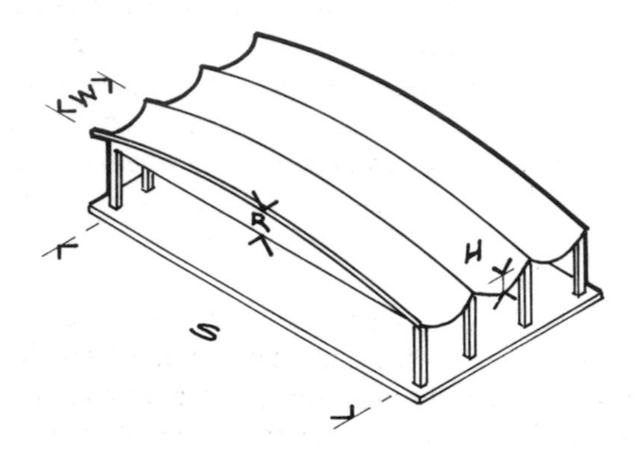

Hyperboloid of Revolution
W about seventh of S
R about same as W
H about twentieth of S

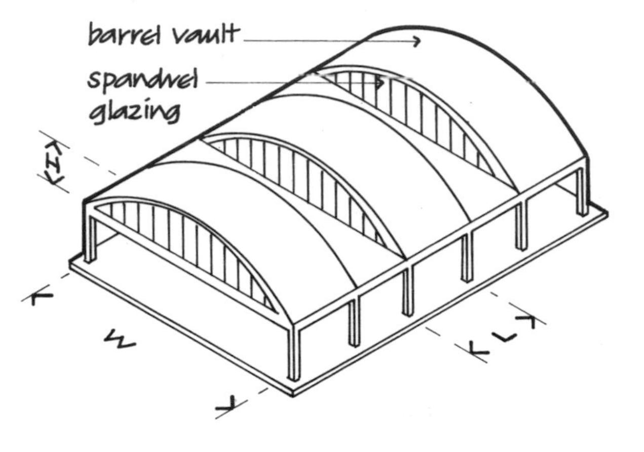

Barrel Vault Shell roof
H about eigth of W
L one fifth of W

Some typical Shell roof forms

Fig. 129

similar vaults. The centering for the conoid, dome and hyperboloid of revolution shells, illustrated in Fig. 129, is considerably more complex and therefore more expensive than that for a barrel vault because of the necessary additional labour and wasteful cutting of material to form support for shapes that are not of a linear uniform curvature.

BARREL VAULT SHELL ROOFS

Reinforced concrete barrel vaults

These consist of a thin membrane of reinforced concrete positively curved in one direction so that the vault acts as structure and roof surface. The concrete

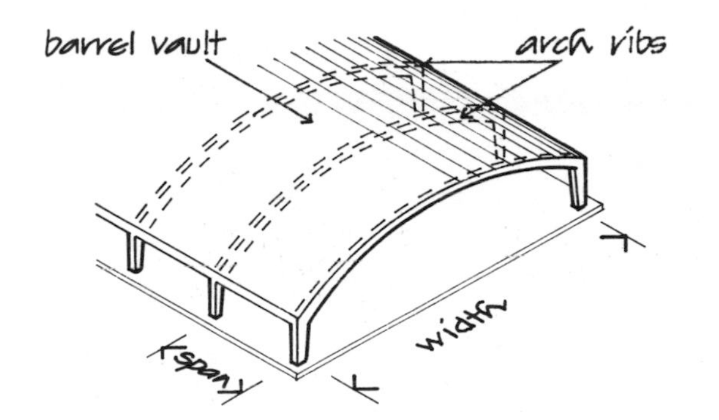

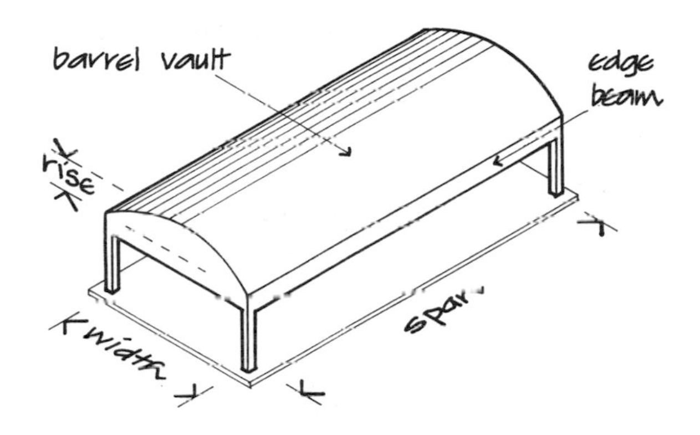

Fig. 130

shell is from 57 to 75 thick for spans of 12.0 to 30.0, respectively. This thickness of concrete provides sufficient cover of concrete to protect the reinforcement against damage by fire and protection against corrosion. The wet concrete is spread over the centering around the reinforcement and compacted by hand to the required thickness. The stiffness of the concrete mix and the reinforcement prevent the concrete from running down the slope of the curvature of the shell while the concrete is wet.

The usual form of barrel vault is the long span vault illustrated in Fig. 130 where the strength and stiffness of the shell lie at right angles to the curvature so that the span is longitudinal to the curvature. The usual span of a long-span barrel vault is from 12.0 to 30.0, with the width being about half the span and the rise about one fifth of the width. To cover large areas, multi-span, multi-bay barrel vault roofs can be used where the roof is extended across the width of the vaults as a multi-bay roof as illustrated in Fig. 131 or as a multi-bay, multi-span roof.

Stiffening beams and arches

Under local loads the thin shell of the barrel vault will tend to distort and lose shape and, if this distortion were of sufficient magnitude, the resultant increase in local stress would cause the shell to progressively collapse. To strengthen the shell against this possibility, stiffening beams or arches are cast integrally with the shell.

Figure 132 illustrates the four types of stiffening members generally used, common practice being to provide a stiffening member between the columns supporting the shell, that is at the limits of the span of the barrel vault. The downstand reinforced concrete beam, which is usually 150 or 225 thick, is the most efficient of the four because of its depth. To avoid the interruption of the line of the soffit of the vaults caused by a downstand beam, an upstand beam is sometimes used. The disadvantage of an upstand beam is that it breaks up the line of the roof and needs protection against weather.

Arch ribs for stiffening barrel vaults, which are less efficient structurally because they usually have less depth than beams, are sometimes preferred for appearance as they follow the curve of the shell and therefore do not appear to interrupt the line of the vault as do beams. The spacing of arch ribs is the same as for beams.

Edge and valley beams

Due to self-weight and imposed loads the thin shell will tend to spread and its curvature flatten out. To resist this, reinforced concrete edge beams are cast between columns as an integral part of the shell. Edge beams may be cast as dropped beams or upstand beams or partly upstand and partly dropped beams, as illustrated in Fig. 133. The advantage of the dropped beam, illustrated in Fig. 130, is that it exposes the whole of the outside of the vault to view. This effect would be spoiled if a rainwater gutter were to be fixed. In hot climates, where rainwater rapidly evaporates and it is not practice to use gutters, the dropped beam edge finish is used. In temperate climates an upstand beam is usual to form a drainage channel for rainwater (Fig. 133).

Similarly between multi-bay vaults a downstand or

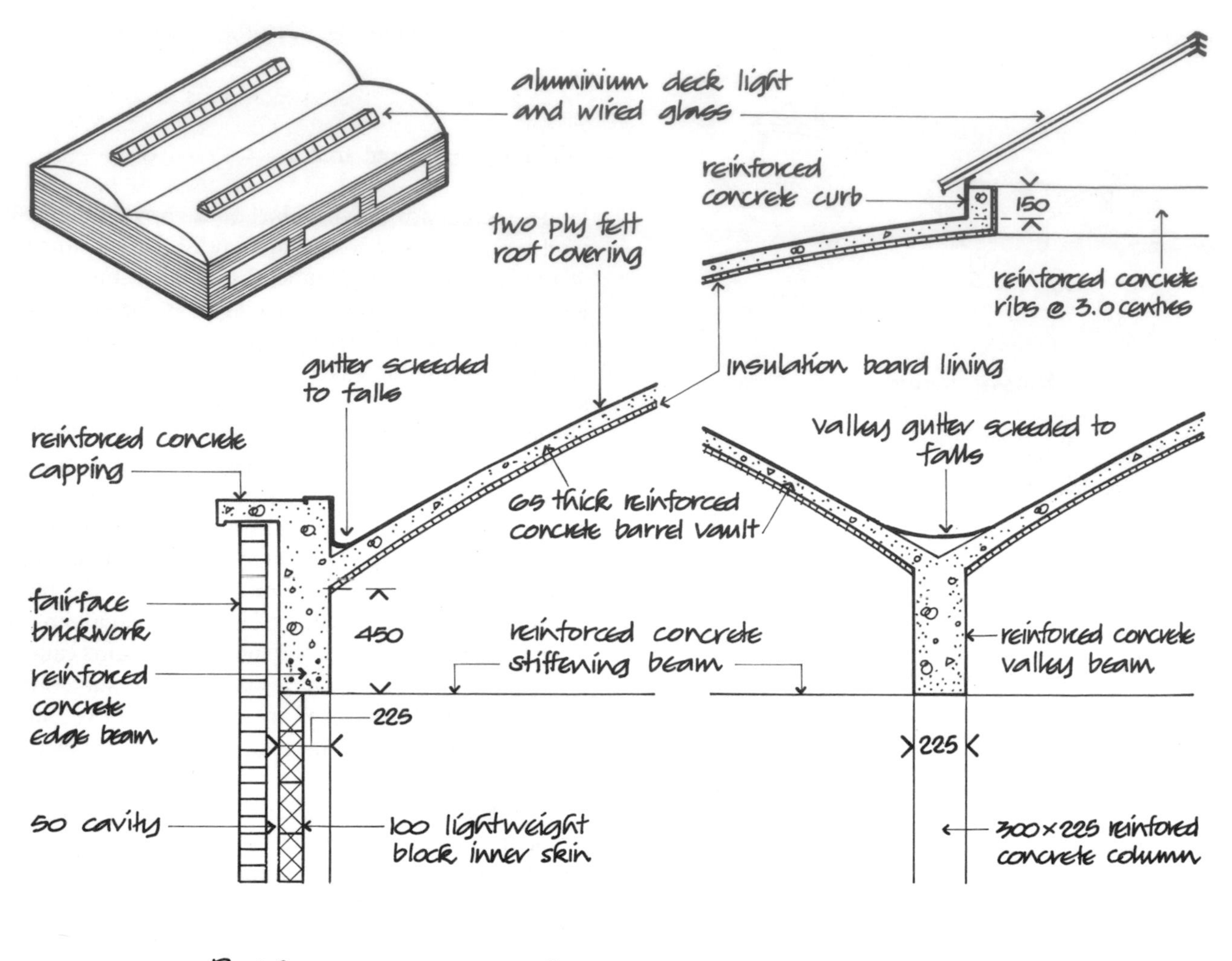

Fig. 131

feather edge valley beam is cast as illustrated in Fig. 131. Spreading of the vaults is largely transmitted to adjacent shells and thence to edge beams on the boundary of roofs, so that comparatively slender feather edge or downstand valley beams are practical.

Rooflights

Top light through the barrel vault can be provided by decklights formed in the crown of the vault, as illustrated in Fig. 131 or by dome lights. The decklight can be continuous along the crown or formed as individual lights. The rooflights are fixed to an upstand curb cast integrally with the shell as illustrated in Fig. 131. One of the advantages of these shells is that their concave soffit reflects and helps to disperse light over the area below. The disadvantage of these top lights is that they may cause overheating and glare in summer months.

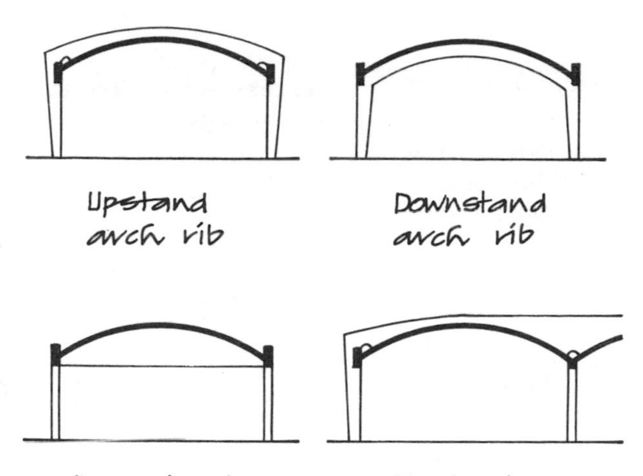

Stiffening beams and arches for reinforced concrete barrel vaults

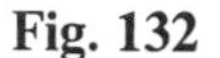

Fig. 132

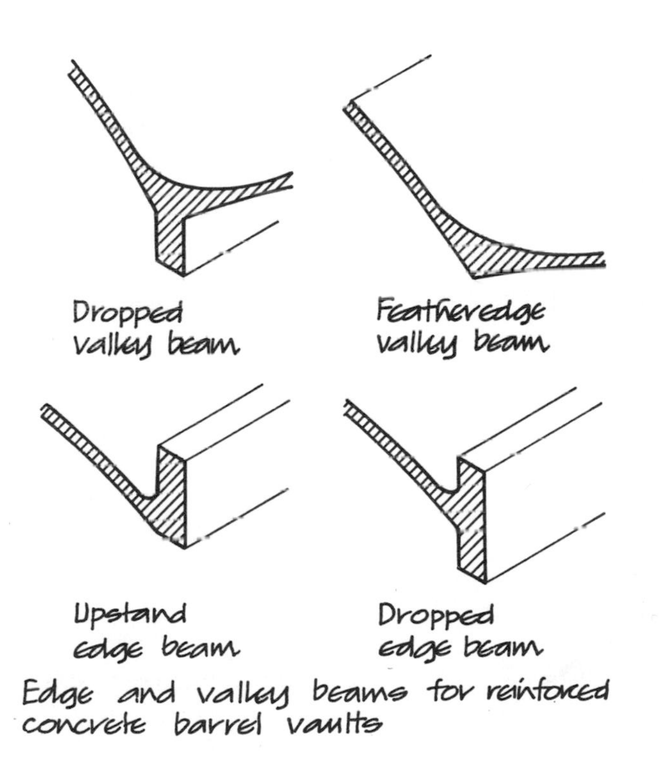

Edge and valley beams for reinforced concrete barrel vaults

Fig. 133

Thermal insulation

The thin concrete shell offers poor resistance to transfer of heat so that some form of insulating soffit lining or a lightweight aggregate screed on the shell may be necessary. The need to add some form of insulating lining to improve insulation adds considerably to the cost of the shell. Pliable insulating boards which can be laid on the centering and take up the curve of the vault will adhere sufficiently to the concrete of the shell to provide adequate fixing. The possibility of condensation forming on the underside of the cold concrete shell and so saturating the insulation makes this an unsatisfactory finish to the soffit of the vault. To fix preformed insulating lining under the vault with a ventilated air space between the shell and the lining would be grossly expensive. The most satisfactory method of providing insulation is to spread a lightweight screed over the shell.

The difficulties of improving the insulation of shells, controlling condensation and at the same time maintaining the elegance of the curved shape of the shell makes these structures largely unsuited to heated buildings in temperate climates.

Expansion joints

With changes in temperature, lineal expansion or contraction of these rigid concrete shells occurs. If there were excessive contraction or expansion the stresses so caused might deform the shell and cause gradual collapse. To limit expansion and contraction, continuous expansion joints are formed at intervals of about 30.0 both along the span and across the width of multi-bay, multi-span barrel vault roofs. In effect the expansion joint is formed by erecting separate shell structures each with its own supports and with compressible and expandable joint material between adjacent structures as illustrated in Fig. 134. The expansion joint transverse to the span of the vaults is formed by casting an upstand to adjacent stiffening beams with a non-ferrous flashing to weather the joint as shown in Fig. 134. The expansion joint is made continuous to the ground with double columns each side of a vertical expansion joint. Longitudinal expansion joints are formed in a valley with upstands weathered with non-ferrous capping over the joint (Fig. 134). This joint is continuous to the ground with a vertical expansion joint between a pair of columns.

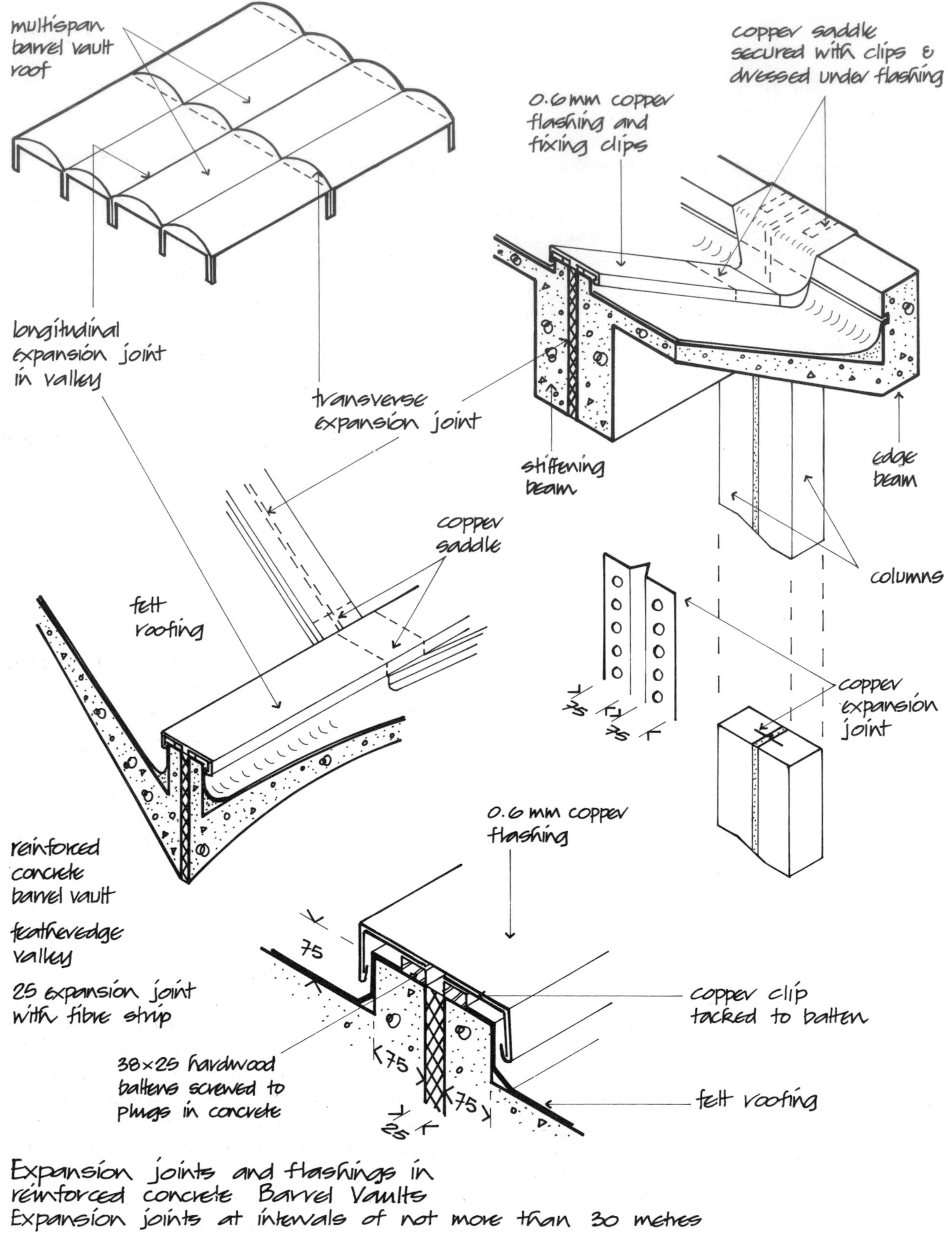

Expansion joints and flashings in reinforced concrete Barrel Vaults
Expansion joints at intervals of not more than 30 metres

Fig. 134

Roof covering

Concrete shells may be covered with non-ferrous sheet metal, asphalt, bitumen felt, a plastic membrane or a liquid rubber-based coating consisting of a neoprene (synthetic rubber) undercoat and chlorosulphonated polyethylene finishing coat applied by brush or spray with reinforcing tape bedded in the material over construction joints in the concrete. This elastomeric coating is supplied in six different colours and being extremely light in weight, that is 0.97 kg/m^2 and resilient, is ideal as a covering for concrete shells. Built-up bituminous felt is often used because it is comparatively light in weight and cheap. Mastic asphalt roofing is a comparatively heavy covering (44 kg/m^2) and is not much used for shell roofs. Non-ferrous sheet metal coverings are fixed to concrete shells in the same way that they are fixed to concrete roofs as described in Volume 1.

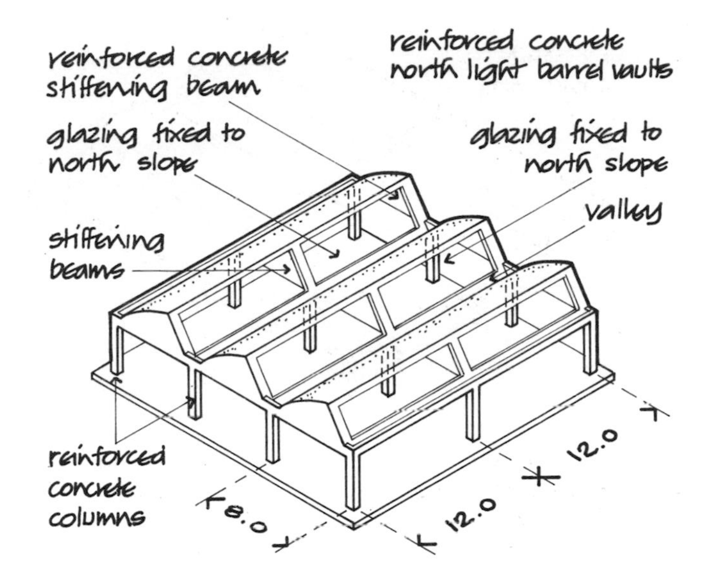

Fig. 135

Walls

The walls of shell structures take the form of non-loadbearing panel walls of brick, block or timber built between or across columns to exclude wind and rain and an insulation against transfer of heat.

North light reinforced concrete barrel vault

To avoid the possibility of overheating and glare from toplights in the summer months a system of north light barrel vaults is used. The roof consists of a thin reinforced concrete shell on the south-facing side of the roof, with a reinforced concrete framed north-facing slope, pitched at from 60° to 80° as illustrated in Fig. 135.

The rigidity of a barrel vault depends on its continuous curvature, which in this type of roof is interrupted by the north light opening. In consequence a north light shell is less efficient structurally than a barrel vault shell. The economic span of a north light shell is 12.0 to 15.0 as compared to the 30.0 or more of the barrel vault.

The reinforced concrete beam and post framing in the north light slope serves as a deep open web beam supporting the crown of the vaulted slope. The north light framing may be open between supporting columns, as illustrated in Fig. 135, or stiffened with intermediate posts as illustrated in Fig. 136. Obviously an increase in the spacing of the posts of the north light frame will require an increase in the section of the eaves and valley beams shown in Fig. 136.

The description of stiffening beams and arches, edge beams, insulation, expansion joints and roof covering given under the heading of barrel vaults applies equally to north light vaults. The north light slope is glazed with patent metal glazing or profiled plastic sheeting fixed to timber grounds or metal angles screwed to the concrete as illustrated in Fig. 136.

Timber barrel vaults

Single- and multi-bay barrel vaults can be constructed from small section timber with spans and widths similar to reinforced concrete barrel vaults. The vault is formed of three layers of boards glued and nailed together and stiffened with ribs at close centres, as shown in Fig. 137. The ribs serve both to stiffen the shell and to maintain the boards' curvature over the vault. Glued laminated edge and valley beams are formed to resist spreading of the vault.

There is no appreciable difference in cost between similar concrete and timber barrel vaults.

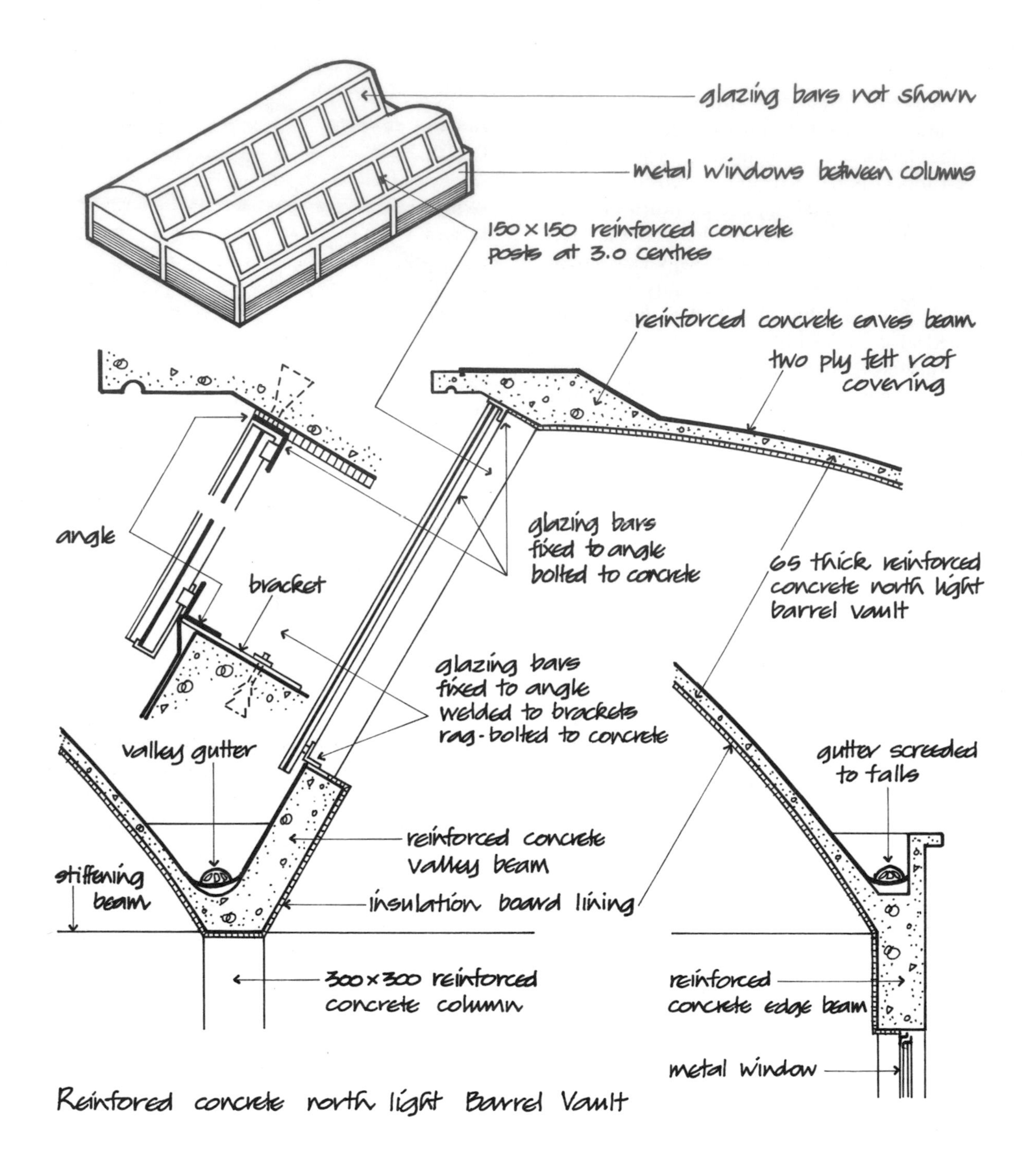
glazing bars not shown
metal windows between columns
150 x 150 reinforced concrete posts at 3.0 centres
reinforced concrete eaves beam
two ply felt roof covering
glazing bars fixed to angle bolted to concrete
angle
bracket
65 thick reinforced concrete north light barrel vault
glazing bars fixed to angle welded to brackets rag-bolted to concrete
valley gutter
gutter screeded to falls
reinforced concrete valley beam
stiffening beam
insulation board lining
300 x 300 reinforced concrete column
reinforced concrete edge beam
metal window
Reinfored concrete north light Barrel Vault

Fig. 136

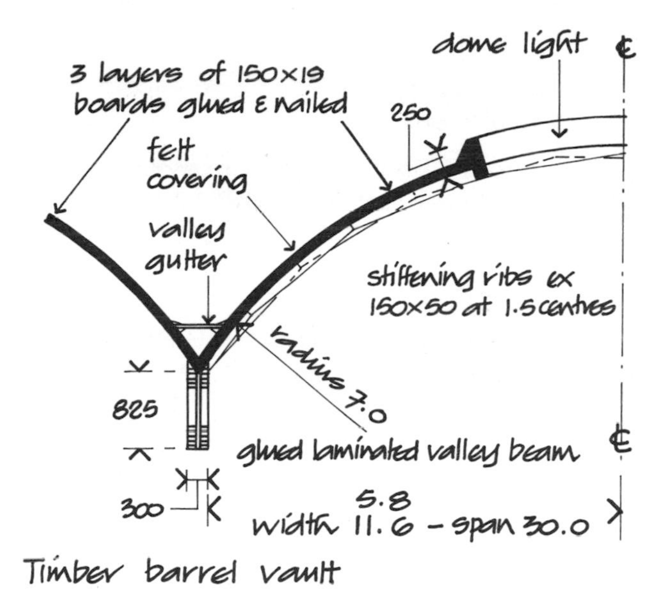

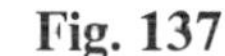
Fig. 137

CONOID AND HYPERBOLOID SHELL ROOFS

Reinforced concrete conoid shell roofs

In this shell form the curvature and rise of the shell increases from a shallow curve to a steeply curved end in which north light spandrel glazing is fixed as illustrated in Fig. 129. The glazed end of each shell consists of a reinforced concrete or steel lattice which serves as a stiffening beam to resist deformation of the shell. Edge beams resist spreading of the shell as previously described.

It will be seen from the illustration of this form in Fig. 129 that, because of the sharply curved glazed end, there is a considerable volume of space inside the shell which cannot be used for production or storage. This particular arrangement of conoid shells is not suitable for use over heated factories and warehouses but is used over long-span enclosures such as railway stations and covered markets where the enclosed space is not heated and a high roof is no disadvantage.

A system of in-situ or pre-cast unit concrete conoid shells with tied lattice steel arches in a range of standard sizes has been used with spans of up to 63.0 and in bays of 7.5, with north light glazing incorporated in the steel arch framing, as illustrated in Fig. 138. This roof system is reasonable in first cost, requires little maintenance and is suited to unheated long-span enclosures.

Hyperbolic paraboloid shells

The hyberbolic paraboloid concrete shells designed and constructed by Felix Candella in Mexico demonstrated the dramatic shapes and structural possibilities of doubly curved shells. This shape is formed when a parabolic generator moves along a parabolic directrix with the plane of the generator remaining vertical as it moves along the directrix (Fig. 139). The resulting surface is described as a hyperbolic paraboloid because horizontal sections through the surface are hyperbolas and vertical sections parabolas.

The structural significance of this shape is that at every point on the surface, straight lines, which lie in the surface, intersect so that in effect the surface is made up of a network of intersecting straight lines. In consequence the centering for a reinforced concrete hyperbolic paraboloid can consist of thin straight sections of timber which are simple to fix and support.

The most usual form of hyperbolic paraboloid roof is a straight line limited section of the shape illustrated in Fig. 140, the form being limited by straight lines for convenience in covering square plan shapes.

To set out stright line limited hyperbolic paraboloid surfaces it is only necesssary to draw horizontal plane squares ABCD and lift one or more corners as illustrated in Fig. 140. The straight lines joining corresponding points on opposite sides set out the surface. The number of lines used to set out the surface is not material except that the more lines used the more clearly the surface will be revealed.

It will be seen from Fig. 140 that this surface is formed by concave downward parabolas running between high points 'a' and 'c' and concave upward parabolas between low points 'B' and 'D'. The amount by which the corners are raised will affect the curvature, shape and strength of the roof. The rise of a straight line limited hyperbolic paraboloid is the difference in height between the high and low points. If three corners are lifted differing heights, then the rise is the mean of the difference between the high and low points.

Obviously if the rise is small there will be little curvature of the shell which will then behave like a plane surface or plate and will need considerable thickness to resist deflection under load. The economic limit of least rise of this shell form is a rise of not less than one fifteenth of the diagonal span, that is the horizontal distance 'AC' in Fig. 140. The greater the rise the less the required thickness of shell.

Straight line limited hyperbolic paraboloids can be combined to provide a structure with rooflights fixed in

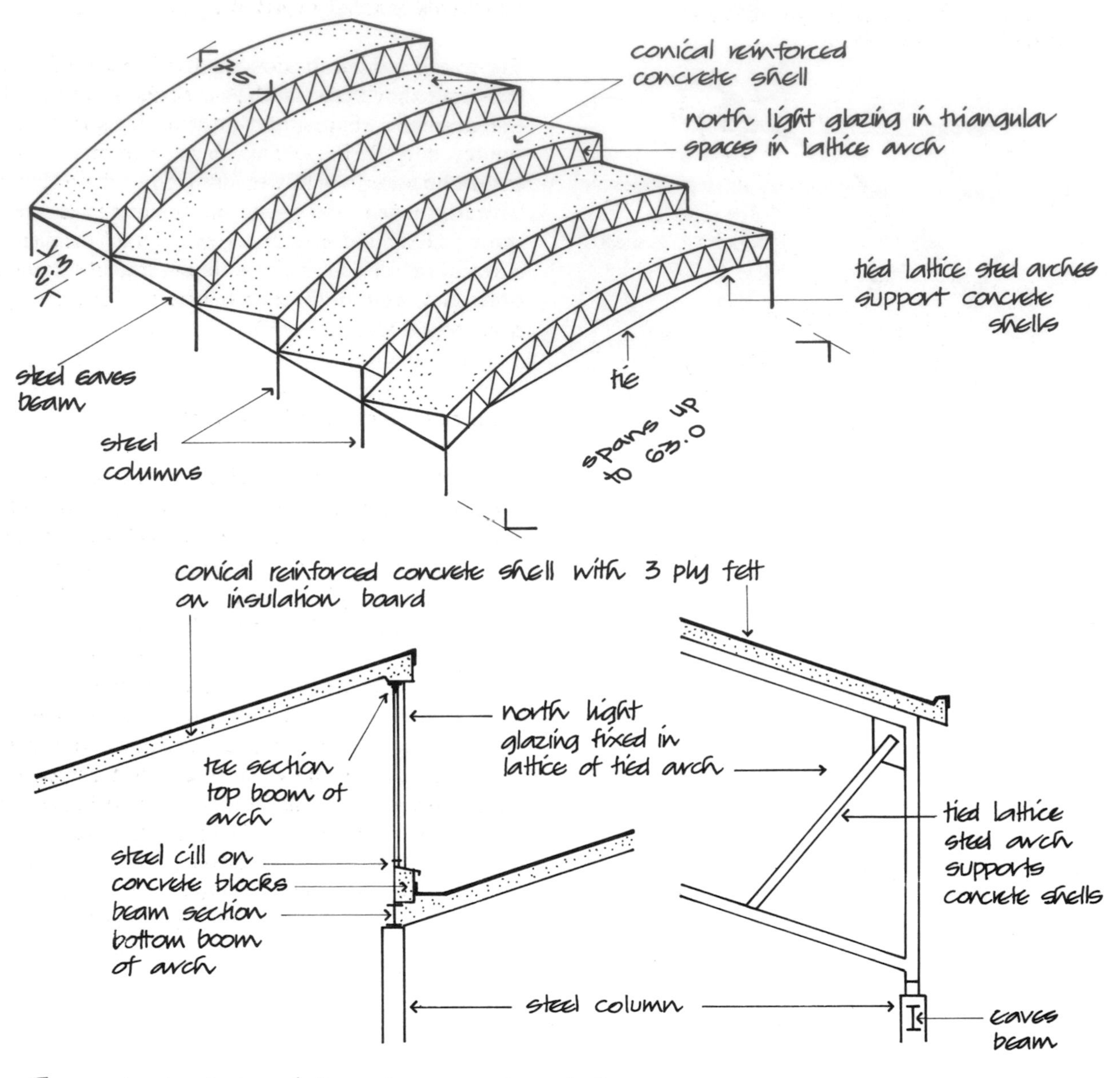
7.5
conical reinforced
concrete shell
north light glazing in triangular
spaces in lattice arch
2.3
tied lattice steel arches
support concrete
shells
steel eaves
beam
tie
steel
columns
spans up
to 63.0
conical reinforced concrete shell with 3 ply felt
on insulation board
north light
glazing fixed in
lattice of tied arch
tee section
top boom of
arch
tied lattice
steel arch
supports
concrete shells
steel cill on
concrete blocks
beam section
bottom boom
of arch
steel column
eaves
beam
Trussed Conical reinforced concrete shell

Fig. 138

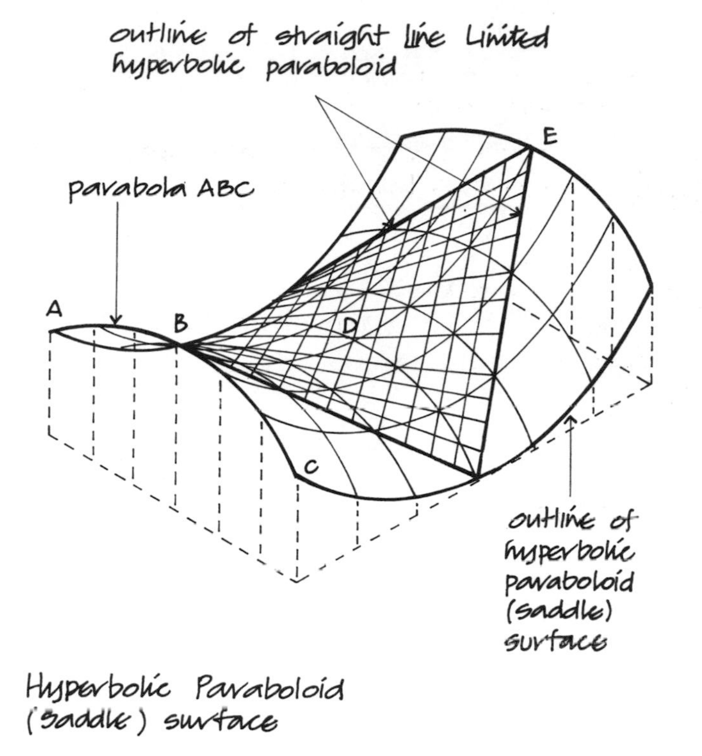

Fig. 139

the spandrel between adjacent shells (Fig. 141), where a number of similar separate hyperbolic paraboloid surfaces are combined to form a roof with spandrel glazing.

Reinforced concrete hyperbolic paraboloid shell

Figure 142 is an illustration of an umbrella roof formed from four hyperbolic paraboloid surfaces supported on one column. The small section reinforcing mesh in the surface of the shell resists tensile and compressive stress and the heavier reinforcement around the edges and between the four hyperbolic paraboloid surfaces resists shear forces developed by the tensile and compressive stress in the shell. A series of these umbrella roofs are combined, with roof glazing between them, to give cover to the floor area below.

Timber hyperbolic paraboloid shell

A hyperbolic paraboloid shell can be formed with three layers of boards nailed together and glued around the edges with laminated edge beams, as illustrated in Fig. 143. It will be seen that the boards do not follow the straight lines lying in the surface of the shell. If they did they would have to be bent along their length and twisted across their width. As it is difficult to twist a board across its width and maintain it in that position, the boards are fixed as shown where they have to be bent only along their length. The timber edge beams, formed by glueing and screwing boards together, resist shear. Low points of the shell are anchored to concrete abutments to prevent the shell spreading under load.

The advantage of a timber shell is its low density of 25 kg/m^2 as compared to the density of 150 kg/m^2 for a similar concrete shell, and the better insulation of the timber.

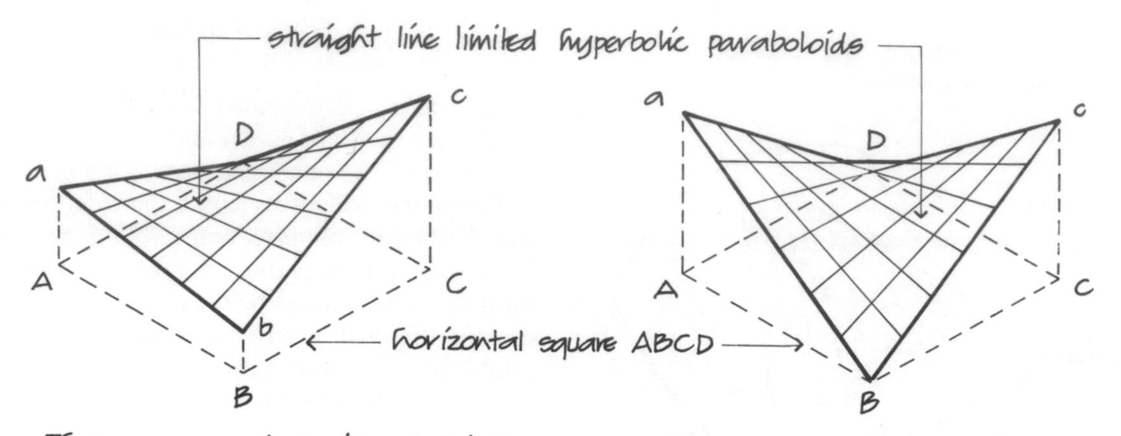

Three corners raised different heights

Two corners raised the same height

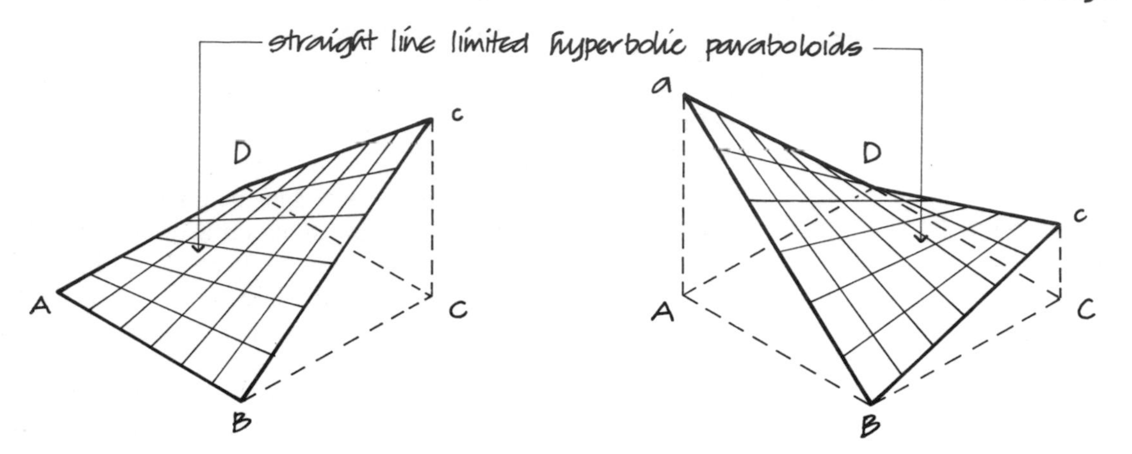

Two corners raised different heights

Setting out straight line limited hyperbolic paraboloid surfaces on a square base

Fig. 140

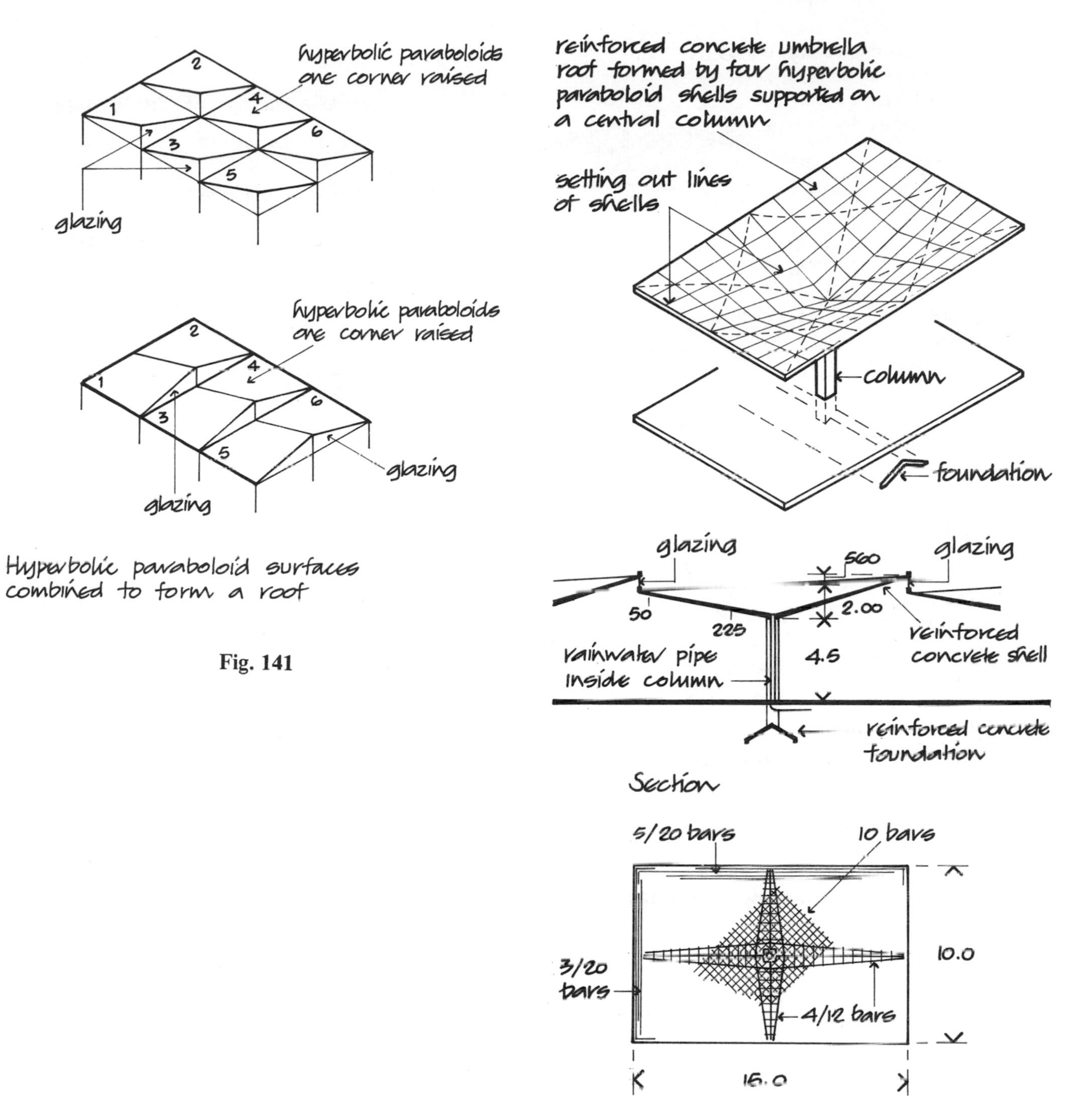

Hyperbolic paraboloid surfaces combined to form a roof

Fig. 141

Reinforced concrete hyperbolic paraboloid

Fig. 142

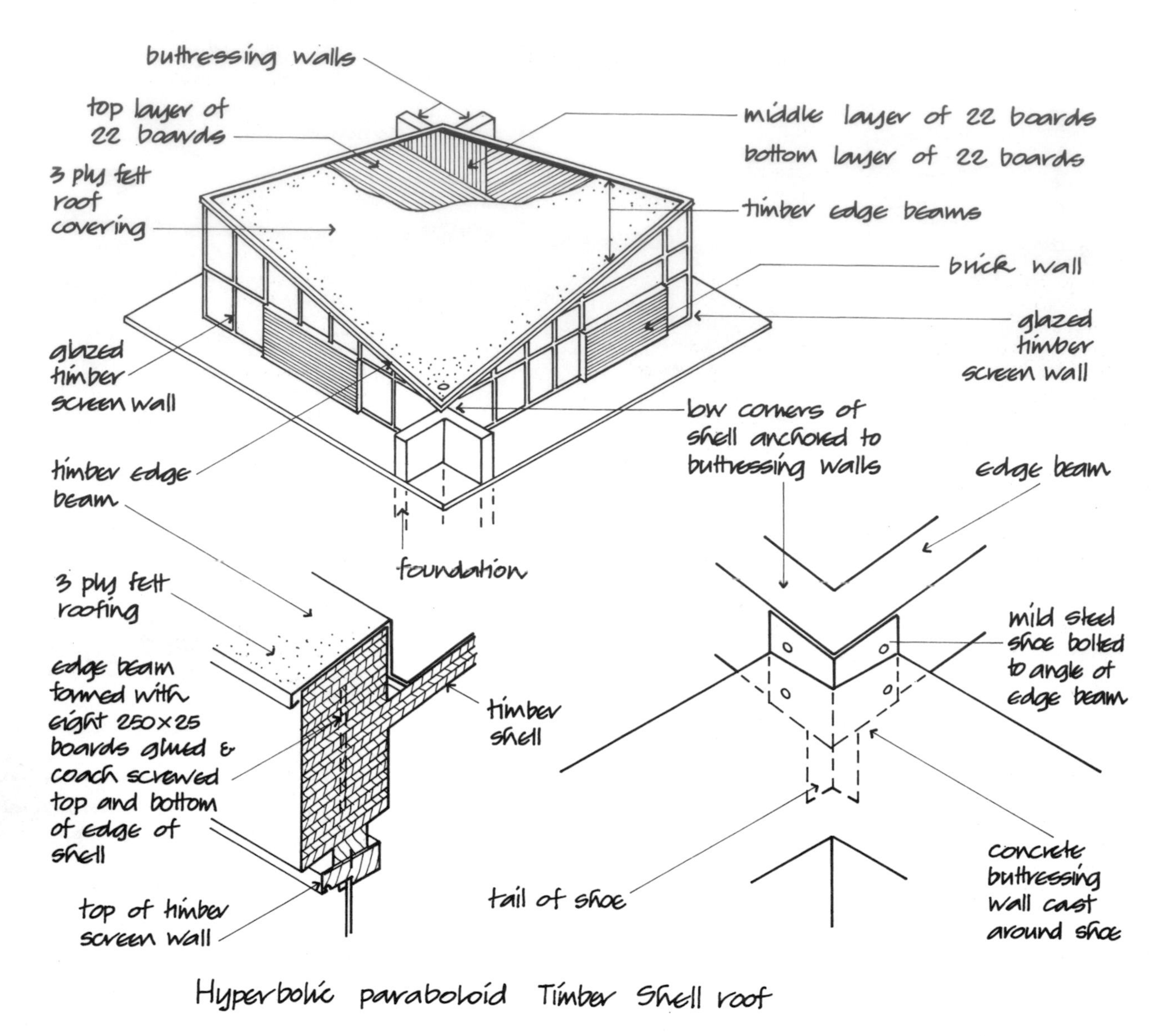

buttressing walls
top layer of 22 boards
middle layer of 22 boards
bottom layer of 22 boards
3 ply felt roof covering
timber edge beams
brick wall
glazed timber screen wall
glazed timber screen wall
low corners of shell anchored to buttressing walls
timber edge beam
foundation
edge beam
3 ply felt roofing
edge beam formed with eight 250×25 boards glued & coach screwed top and bottom of edge of shell
timber shell
mild steel shoe bolted to angle of edge beam
top of timber screen wall
tail of shoe
concrete buttressing wall cast around shoe
Hyperbolic paraboloid Timber Shell roof

Fig. 143

INDEX